BUILDING STONES of DUBLIN

A Walking Guide

Patrick Wyse Jackson

with best wishes,
Patrick Wyse Jackson

Photography by Declan Burke

GEOLOGISTS' ASSOCIATION
1858

COUNTRY
HOUSE

Dedicated to the memory of
John Semple Jackson

Published in 1993 by
Town House and Country House
42 Morehampton Road
Donnybrook
Dublin 4
Ireland

A CIP catalogue record for this book is available from the British Library.

ISBN: 0-946172-32-3

Published with the generous assistance of Stone Developments plc, Enniskerry, Co. Wicklow; Dr Eric Robinson, London; the Department of Geology, Trinity College, Dublin; and the Curry Fund of the Geologists' Association, London.

Front cover: View of Christ Church Cathedral, Dublin
(Photo: Liam Blake)

Text editor: Elaine Campion
Design: Bill Murphy
Colour separations by The Kulor Centre, Dublin
Printed in Ireland by Criterion Press, Dublin

CONTENTS

ACKNOWLEDGEMENTS

Much of the impetus and encouragement for this book came from the late Dr John Jackson. John, who was the most generous of men, gave me his own notes on Irish stone to use, and read the first draft of this book, for which I am most grateful. He enlivened a brief holiday spent in Schull, Co. Cork, with his geological and anecdotal talk, and through his and Sally's warm hospitality.

I am particularly indebted to the Curry Fund of the Geologists' Association, London; Stone Developments plc, Co. Wicklow; Dr Eric Robinson, London; and the Department of Geology, Trinity College, Dublin, for their generous assistance towards publication costs.

I am also grateful to the following members of the stone trade, who answered my questions concerning building stone: Richard Lyons (Castone Supplies), Barry Feely (Feely and Sons, Roscommon), Michael Kirwan (Interclean), Tessie Joyce (Connemara Marble Industries Ltd, Moycullen), Thomas Murphy (James Murphy and Co., Sandyford), Derry O'Sullivan (Roadstone Exports Ltd, Arklow), Catherine Roe (Roe and O'Neill, Sandyford), Tom Hayden and Maura Murray (Capco Ltd, Terenure), and Michael O'Connor and John Donohue (Stone Developments, Enniskerry). Many kindly supplied stone samples, which have been deposited in the Geological Museum, Trinity College, Dublin.

I have received information during the course of this work from a large number of people. I thank Maeve Boland, Petra Coffey, John Feehan, Aubrey Flegg, Russ Heselden, Dave Johnston, Eddie McParland, Nigel Monaghan, Joan O'Connor (New Ireland Assurance Co.), Eric Rankin (McDonnell & Dixon, Architects) and Ian Sanders.

I thank all the Wyse Jackson family for their encouragement, especially Vanessa, who has walked much of Dublin with me.

Declan Burke gratefully acknowledges advice and the use of facilities at the Department of Photography, College of Technology, Kevin Street, Dublin.

Patrick Wyse Jackson

INTRODUCTION

Grey brick upon brick,
Declamatory bronze
On sombre pedestals —
O'Connell, Grattan, Moore —

Louis MacNeice — Dublin

MacNeice exemplifies grey brick with an image of a dark
sombre Dublin. Close examination of Dublin's buildings
reveals no grey brick, rather yellows and reds prevail.
Perhaps MacNeice was attempting to conjure up the image
immortalised in the sobriquet 'Dear Dirty Dublin'. This dirt
and darkness that consumed Dublin in the 1800s and early
1900s was due to two factors: the burning of coal, and the
use of a dark muddy limestone called Calp in buildings
and for paving. However, it is brick and not Calp that to
MacNeice's readers would characterise Dublin.

As Dubliners, we tend to live our lives and make our
arrangements according to landmarks. As the city evolves
many of these sites fall before the bulldozer, yet they are
still vividly fresh in our mind's landscape. Tea is still
available at the DBC*; wonderful exotic foods at Smyth's on
the Green; buses still pass by the Nelson Pillar; and
browsers are still leafing through the APCK[†].

In contrast, many landmarks are still standing, but we,
like MacNeice, fail to recognise their features clearly. A
friend of mine refers to the church at the top of Foxrock
Avenue as the 'red brick church'. Red brick? No! It's faced
in fresh grey granite from Larch Hill, Co. Dublin!

* *Dublin Bread Company*
[†] *Association for the Promotion of Christian Knowledge*

This book describes stone and other building materials encountered on four walks in the city centre. It includes a glossary of architectural and geological terms as well as charts that classify different rock types. The text for each walk is accompanied by a map, and sixteen of the stone types are illustrated in colour.

The book seeks to guide the walking reader through parts of the city, some medieval, some new, and to highlight the use of a variety of native and exotic building stones. These materials comprise a portion of our heritage that has been neglected and deserves attention. If this guide reduces the number of 'red brick churches' in the cloisters of the reader's mind it will have fulfilled its purpose.

Patrick Wyse Jackson
Dublin: 28 January 1993

THE USE OF STONE AND OTHER BUILDING MATERIALS IN IRELAND

Stone is one of the most versatile building materials. It can be cut, broken, sawn, drilled and carved. The stone most suitable for building is the type that may be quarried in large blocks, which have no blemishes, cracks or other flaws. Such stone is called dimension stone. Stone is strong, durable and in many cases is resistant to breakdown, so that buildings constructed of it generally have a long life.

The use of stone in Ireland reflects the geology (underlying rock types) (Fig 1), and for the most part local stone was used in the past, when transportation was difficult. Broadly, rocks may be divided into three convenient groupings that reflect their origins.

Igneous rocks are those that are derived from molten magma deep in the earth (Fig 2). They may appear on the surface through a volcanic eruption or, as in the case of granite, they may rise slowly below the crust. The mass of molten rock that forms granite cooled and crystallised slowly, and was only exposed at the surface after the erosion of the overlying layers of rock. Granite occurs in a variety of colours and textures, and is a tough crystalline rock composed of three minerals — quartz, mica, and feldspar. Leinster Granite has been used for a long time in the Dublin area, and is quarried today at Ballybrew, near Enniskerry, and at Ballyknockan, near Blessington, both in Co. Wicklow. An excellent account of the history of quarrying in the Three Rock area of the Dublin Mountains is given by Nicholas Ryan in his book *Sparkling Granite* (1992). The Leinster Granite was intruded in a number of distinct units which differ slightly in mineralogy and texture (see Brück and O'Connor, 1977, for more information). Granites have also been exploited in Counties Galway, Donegal and Down. The granite is removed in large

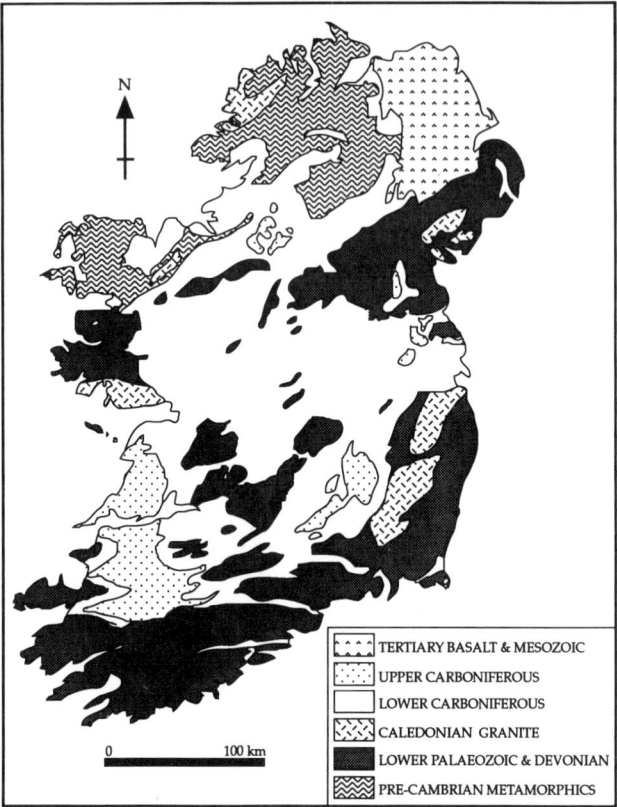

Fig 1   Simplified geological map of Ireland.

Mineral content / Texture	Quartz Feldspar Biotite Hornblende	Feldspar Biotite Pyroxene Hornblende	Feldspar Pyroxene Olivine
Coarse-grained	Granite Granodiorite	Diorite	Gabbro
Medium-grained	Microgranite		Dolerite
Fine-grained	Rhyolite	Andesite	Basalt
Colour	Pale ⟶		Dark
% silica	[acidic] 65	50	[basic] 45

Fig 2   Classification of igneous rocks. They are differentiated by texture and mineral content.

blocks by using low-yield explosives or diamond-studded wire, and then sawn and finished before being fitted to buildings by metal ties.

Sedimentary rocks are those that are derived from the weathering and erosion of pre-existing rocks (Fig 3). The resultant products are transported some distance before being deposited in layers (beds or strata) in environments as diverse as rivers, deserts or marginal marine areas.

CLASTIC (INORGANIC) SEDIMENTS (composed of cemented fragments of rock; classified according to size of the fragments).

RUDACEOUS (Cobbles: >2mm) ⟨ BRECCIA (angular) / CONGLOMERATE (rounded)
ARENACEOUS (Sand: $2 - \frac{1}{16}$ mm) - SANDSTONE
ARGILLACEOUS (Silt: $\frac{1}{16} - \frac{1}{256}$ mm) - SILTSTONE; SHALE; GREYWACKE

SEDIMENTS OF CHEMICAL ORIGIN (formed by precipitation from solution, or by replacement of one mineral by another).

CALCAREOUS (Limy) - OOLITIC LIMESTONE ($CaCO_3$);
TRAVERTINE; DOLOMITE ($MgCO_3$)
SILICEOUS (Composed of silica) - FLINT (pale); CHERT (dark)
FERRUGINOUS - IRONSTONE
EVAPORITES (Saline) - ROCK SALT

SEDIMENTS OF ORGANIC ORIGIN (formed by accumulation of plant or animal organic matter).

CALCAREOUS (Limy $CaCO_3$) - LIMESTONE
CARBONACEOUS (rich in carbon) - COAL
PHOSPHATIC - BONEBED (PHOSPHATIC ROCK)

Fig 3   Classification of sedimentary rocks. They are largely determined by the type, nature and size of the sediment.

In Ireland, *sedimentary rocks* include Old Red Sandstone, which makes up the mountains of Cork and Kerry. This clastic rock, composed of sand grains cemented together with iron oxide, was commonly used for building in the south-west of the country. The Old Red Sandstone was deposited in a variety of environments: in river channels, in deserts as dunes, or in shallow marine areas. Another sedimentary rock is the well-known Liscannor Flagstone of Co. Clare, which is composed of silts. These silts contain burrows or trace fossils, which are indicative of animal activity. The stone may be split easily into flat pieces, and is usually used for paving.

The most common sedimentary rock in Ireland is limestone, which underlies the Central Plain and most of Dublin. It is usually grey and fossiliferous, and is quarried in Counties Galway, Kilkenny, Carlow and Roscommon.

Limestones that take a polish are referred to as marbles, but in strict geological terminology this is incorrect. A true marble is a *metamorphosed rock* that has undergone physical change. Several types of metamorphism are recognised (Fig 4). Contact metamorphism occurs when a rock is altered by heat from an adjacent igneous intrusion. This has occurred in a belt one kilometre wide around the Leinster Granite, in which the Lower Palaeozoic sediments have been altered to schists. More wide-scale alteration is called regional metamorphism and is caused by heat and pressures generated when cataclysmic events, such as continents colliding, take place. This results in rocks at the margins of the continents being severely folded and uplifted to form mountains. These events, called orogenies, have happened several times in the past. The Himalayas and Alps were formed in this way approximately 20 to 50 million years before the present, as were the mountains of Donegal and Scotland between 400 and 500 million years ago.

Metamorphic type / Original rock type	Contact 700-1000°C	Regional		
		Low grade	Medium grade	High grade
Limestone (pure)	Marble	Marble	Marble	Marble
Shale/Mudstone	Hornfels	Slate/Phyllite	Schist	Gneiss
Sandstone	Quartzite	Quartz Schist	Quartzite	Quartzite
Greywacke		Schist	Schist	Gneiss
Basalt	Basic Hornfels	Greenschist	Amphibolite	Amphibolite

Fig 4   Classification of metamorphic rocks. The type of rock produced is determined by the original material and the degree of alteration to which it has been subjected.

During metamorphism, limestone is converted into marble, shales into slates, and sandstones into quartzites. In Ireland, the metamorphic rocks most frequently exploited were slates from Counties Tipperary and Kerry, and the distinct and decorative Connemara Marble. This rock, which originally was a lime mud, was deposited in a shallow sea 600 million years ago, and was altered by heat and pressure approximately 500 million years ago; new metamorphic minerals, such as serpentine and chlorite, developed as the rock equilibrated with the new conditions to which it had been subjected.

The earth is dynamic, and since its formation some 4600 million years ago it has been constantly changing: continents are waxing and waning, ocean basins are opening, volcanoes are erupting, and rocks of different types are forming. Geologists have for convenience divided geological time into periods, of which the geological period is the most useful (Fig 5). In the British Isles, each period was characterised by different environmental conditions, which affected the type of rocks produced in that time. Thus, during the Carboniferous period, when shallow seas covered most of these islands, limestone was deposited. Later, during the late Permian and Triassic periods, arid desert conditions prevailed and sandstones were deposited.

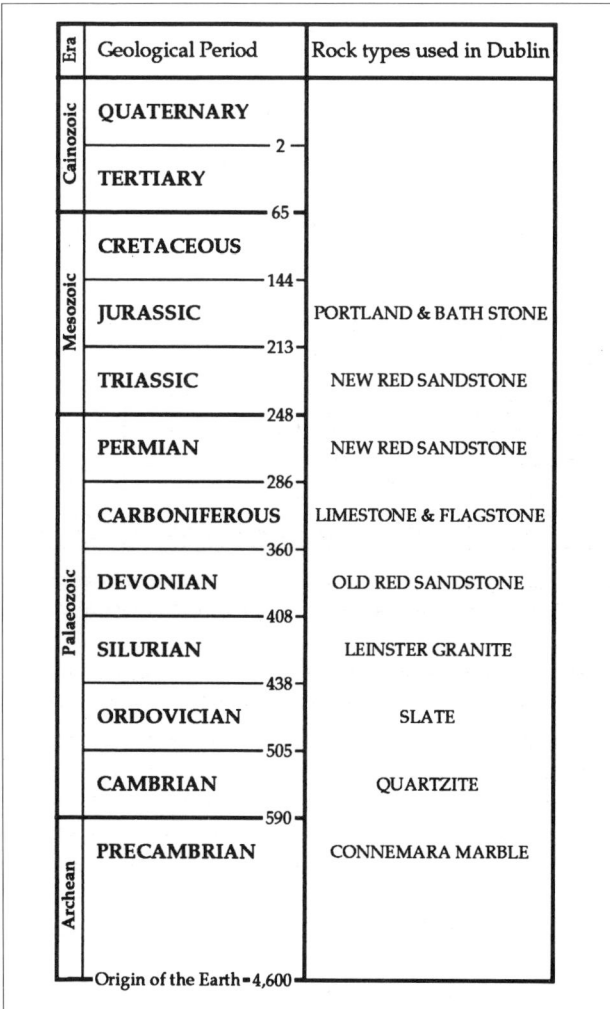

Era	Geological Period	Rock types used in Dublin
Cainozoic	QUATERNARY	
	— 2 —	
	TERTIARY	
	— 65 —	
Mesozoic	CRETACEOUS	
	— 144 —	
	JURASSIC	PORTLAND & BATH STONE
	— 213 —	
	TRIASSIC	NEW RED SANDSTONE
	— 248 —	
Palaeozoic	PERMIAN	NEW RED SANDSTONE
	— 286 —	
	CARBONIFEROUS	LIMESTONE & FLAGSTONE
	— 360 —	
	DEVONIAN	OLD RED SANDSTONE
	— 408 —	
	SILURIAN	LEINSTER GRANITE
	— 438 —	
	ORDOVICIAN	SLATE
	— 505 —	
	CAMBRIAN	QUARTZITE
	— 590 —	
Archean	PRECAMBRIAN	CONNEMARA MARBLE
	Origin of the Earth = 4,600	

Fig 5   Geological timescale and stone types used in Dublin, diagnostic of particular geological periods. Ages are given in millions of years before the present.

Stone may deteriorate through the action of atmospheric pollution, percolating water, growth of algae, alternating cycles of heat and cold, rusting of metal ties that hold stone blocks together, or in certain atmospheres through the reaction of its constituents with polluted air. In many cases stone breaks up through the precipitation of salts within the fabric of the rock. This has happened on the National Museum and the National Library in Kildare Street, Dublin. The Mount Charles Sandstone used to face the buildings broke down within forty years and has been replaced on the National Library with Ardbraccan Limestone.

Stone may be cleaned in a variety of ways. Washing with water causes least damage, but is slow. Other methods include sand-blasting the stone surface with grit, which knocks off the dirt but which may remove a great deal of stone. Acids are often used on granites and alkaline materials on limestones, but these substances may also remove much surface rock. Where it is desirable to save architectural and finely carved details, cleaning can take a long time. In such instances the stone is simply washed, or a poultice is applied to the stone which removes some dirt. The rest is etched off using tiny engraving tools.

Stone gives urban Ireland a regional character, and this is reflected in the present urban landscape. Many of Belfast's buildings used basalt from Antrim and granite from the Mourne Mountains. Granite from the Wicklow and Dublin Mountains, and limestone from the immediate hinterland, were used for the same purposes in Dublin. While one can recognise these stone types as characteristic to Dublin, this book points to over one hundred other varieties that are used in Ireland's capital city — including many quarried in Ireland (Fig 6).

Pagan Ireland

The use of stone for building in Ireland goes back at least 5000 years. The pre-Christian inhabitants used local stone to build burial places: dolmens, such as those in the Burren, Co. Clare, and passage graves, including Newgrange, Co. Meath, which is 4500 years old. They also built stone forts. Staigue Fort, near Sneem, Co. Kerry, is one of the best examples. It was built some 2000 years ago of local sandstone. Stone for these structures was always derived locally. In the main it is unlikely that early builders moved stone large distances, although granite cobbles for Newgrange were transported over thirty kilometres. The earliest known quarries are in Co. Sligo and have been dated from 2500 BC.

Plate 1 IGNEOUS ROCKS

a. Ballyknockan Granite
b. Balmoral Red Granite
c. Pink Porrino Granite
d. Blue Pearl Larvikite

a. b.

c. d.

Plate 2 SEDIMENTARY ROCKS

a. Ballinasloe Limestone
b. Calp Limestone
c. Clonony Brown Marble
d. Little Island Cork Red Marble

a. b.

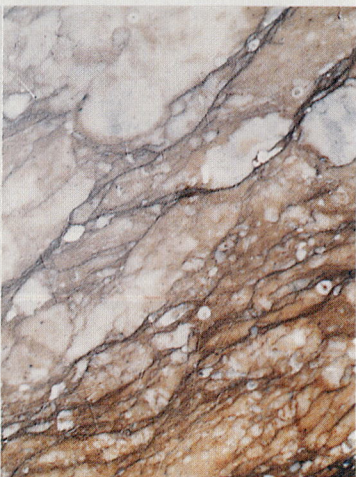

c. d.

Plate 3 SEDIMENTARY ROCKS

a. Liscannor Flagstone
b. Portland Stone
c. Roman Travertine
d. Liverpool Stone

a. b.

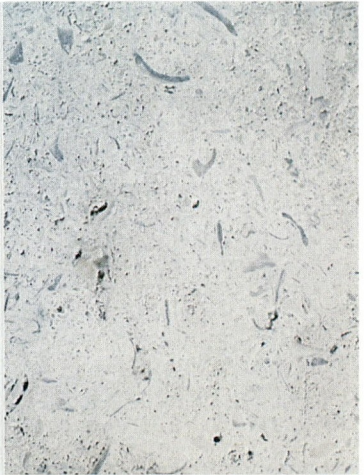

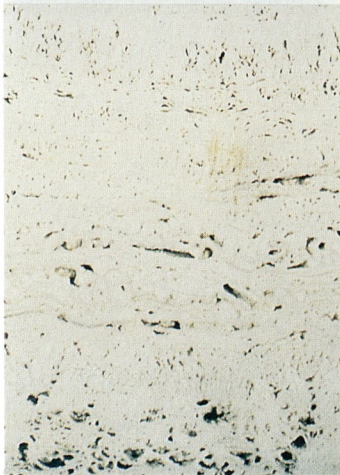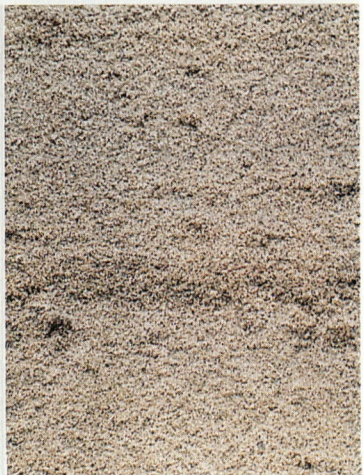

c. d.

Plate 4 METAMORPHIC ROCKS

a. Connemara Marble (Barnaoran)
b. Donegal Quartzite
c. Maracana Gneiss
d. Westmoreland Green Slate

a. b.

c. d.

*Fig 6 Map showing the location of
Irish stone mentioned in the text.*

Scawt Hill
(Scaughthill)

Mount Charles

Armagh

Castle Caldwell

Mourne Mountains O

Cavan

Sheephouse

Ardbraccan

Streamstown
Clifden
Barnaoran
Ballinahinch
Clonmacnoise
Donnybrook
Barna O
Galway
Ballinasloe
Lucan/Palmerstown
Dalkey
Tullamore
Barnacullia
Golden Ball
Clonony
Ballybrew
Blessington
Golden Hill
Liscannor / Doonagore
Ballyknockan
Ashford
Aughrim
Killaloe
Old Leighlin
Parnell Quarry
Limerick
Paulstown
Pallaskenry
Kilkenny

Mitchelstown
Doneraile

Valentia Island
Little Island
Midleton & Baneshane
Churchtown

O GRANITE QUARRIES
● LIMESTONE QUARRIES
✚ OTHER STONE TYPES

The Early Christian period

Early Christian churches were simple, like the seventh-century oratory at Gallarus in Co. Kerry, where no mortar was used. Mortar as a binding agent was introduced later, and the earliest records of its use in Ireland date to the ninth and tenth centuries, when it was used in the construction of round towers. In the eleventh century mortar was used in the construction of churches. During the Early Christian period, differential use of stone types appeared: local stone was used for the bulk of building, and more exotic material was used for architectural features such as carved architraves around windows. Just such use is well developed in St Brendan's Cathedral, Ardfert, Co. Kerry.

Viking, Norman and medieval Dublin

The use of stone for building in Dublin goes back a long time; the first major users were the Normans, who employed both native Calp Limestone and English oolitic limestone from Dundry, Somerset, to build Christ Church Cathedral in 1170.

Calp Limestone was the dominant and cheap building stone in Dublin from Norman times until the late 1700s. It is a very dark, muddy, well-bedded limestone, which contains thin layers of black siliceous chert, and was quarried at Crumlin and Rathgar in Dublin in medieval times. The City Wall, best seen at Great Ship Street and St Audoen's Arch, was built of Calp Limestone. The nickname 'Dear Dirty Dublin', coined by Lady Morgan (Sydney Owenson), came from the use of poor-quality Calp Limestone for paving and street construction. It was quarried from where Donnybrook Bus Station now stands. When this limestone weathered, it released a good deal of mud, which ended up on the streets.

The early eighteenth century

During the eighteenth century, probably Dublin's most prosperous period, most of Ireland's important public buildings were erected. Many of those in Dublin, such as the General Post Office (GPO), the Parliament House (Bank of Ireland), and the Custom House, were built of Calp Limestone rubble walls which were faced with either Leinster Granite or Portland Stone. The latter is an oolitic limestone quarried in Dorset on the south coast of England, and it was often used for the finer features such as swags, columns and statues. This pale limestone was deposited in a shallow sea 150 million years ago, during the geological period known as the Jurassic.

One of the last major buildings to be constructed largely of Calp was the Old Library in Trinity College, erected in the first half of the 1700s.

Granite was available from a large number of locations in the Dublin hinterland: Dalkey, Three Rock Mountain, Kilgobbin, Golden Ball, and Blessington, among others. Apart from building construction, the stone was also used in many engineering works, such as Dún Laoghaire Harbour (Dalkey Granite), and for paving and kerb stones, of which there are still many examples in the city centre. It was used in the Dublin to Kingstown (Dún Laoghaire) Railway for railway sleepers, walls and buildings, and was obtained from temporary quarries close to the line of the railway. The granite sleepers proved most unsuitable and kept breaking, and soon were replaced by timber. Some survive, however, being built into walls adjoining the railway near Booterstown and Blackrock stations.

Portland Stone became very popular through the example of Christopher Wren, who used it in the rebuilding of London after the Great Fire of 1666. Soon it became fashionable for the aristocracy to build their great houses of the stone, and so it developed into a major social and prestigious symbol. Indeed, many older houses in Britain were refaced with Portland Stone at great expense to their owners. It was the ideal medium for stonemasons to carve their elaborate designs. The stone was imported into Dublin in large quantities; the transportation costs involved were less than those to Birmingham!

Artificial stone was introduced to Dublin in the 1750s, and was favoured by the architect James Gandon. One successful variety was Coade Stone, a tough ceramic material of kaolin clay, grog (crushed pottery), flint, and glass (see Freestone, 1991). Produced in London, it could be cast into a variety of intricate shapes, panels and friezes. It was used on the Rutland Fountain opposite the National Gallery of Ireland, and for the frieze of swags and ox-heads on the Rotunda Hospital (see Ruch, 1970, and Kelly, 1990). Examination of Coade pieces shows that the material has decayed very little in nearly 250 years.

Brick has a long history, being used first in Ireland for Carrickfergus Castle, Co. Antrim, in 1575. It became popular during the reign of Queen Anne, and in Dublin it was used extensively from the beginning of the eighteenth century in the great residential squares, such as Fitzwilliam, Mountjoy and Merrion Squares. Most of this brick was imported from England, but some came from brickfields in the inner city or from more distant places such as Athy, Co. Kildare.

Slate for roofing was largely Welsh, although Irish slate was available from Ashford, Co. Wicklow, Killaloe, Co. Clare, and Valentia Island, Co. Kerry. These Irish slate quarries employed at least a thousand men and boys; in 1836 the output at Killaloe was 10,000 tons (10,160 tonnes). The slates tended to be quite thick, and thus were less desirable for building than the thinner Welsh varieties. However, apart from roofing, the Valentia slates were used for billiard tables, worktops and shelving, as they could be split into very large pieces. Some good examples of this slate are to be found in the Public Record Office in London, where it is used for miles of shelving, and on the roofs of houses in North Great George's Street in Dublin.

In the early 1800s, some 12,000 tons (12,192 tonnes) of slate per annum were imported into Dublin. In 1896, a strike at the Welsh quarries led to the Irish market being temporarily flooded with American slate. Irish slate could not compete with the better-quality Welsh slates, which were in any case cheaper to transport to Dublin. The imposition of a £5 per ton tariff on Welsh imports during the 1930s saw a brief revival in Irish slate production, which began to decline again in the 1940s. The quarry at Killoran, Co. Tipperary, which produced 'Killaloe' slates, closed in 1953, but encouragingly for the Irish stone trade, it has recently reopened, and is producing good-quality stone.

Mid nineteenth to early twentieth centuries
The flamboyance of the Victorians is reflected in their use of stone on contemporary buildings. It was fashionable to use bright, colourful stone, and little expense was spared in getting it. Muckross House, erected in 1840 for the Herbert family of Killarney, Co. Kerry, was built of orange-coloured New Red Sandstone which was imported from Chester in England and transported to the site from Kenmare by horse and cart.

The builders of the day made extensive use of imported Scottish and English sandstones, French and English oolitic limestones, Irish limestones, and granite from Wicklow and Dublin. Granite from Castlewellan in the Mourne Mountains was particularly popular in London, where hundreds of tonnes of it were used in the Albert Memorial.

Irish marble was often used as a decorative stone in buildings dating from the early seventeenth century, and its use reached a peak in the mid nineteenth century. The chief marbles included the Connemara green serpentinite marble, the red Cork 'marbles' and the black Kilkenny and Galway 'marbles'. The latter three 'marbles' are not true marbles, being merely limestones capable of taking a good polish. Many quarries were in operation in the mid 1800s. The largest was probably at Little Island, just outside Cork city, which produced 56,000 tons (56,896 tonnes) of 'Victoria' red marble in 1836. The marble industry thrived until the 1890s, after which it slowly declined. Today, Connemara Marble is used almost solely for souvenirs, while the black varieties of Galway, Carlow and Kilkenny are still available and widely used in the construction industry.

Where money was not available to face a building completely in stone, brick was substituted, and the windows and other features were decorated with cut and carved stone. The suburbs of Rathmines and Donnybrook were largely constructed of bricks imported from Bridgewater in England or transported via the Grand Canal from Athy, Co. Kildare. The Victorians also employed specially shaped bricks and coloured clay terracotta tiles to embellish their buildings. Fine examples of terracotta in Dublin include Baggot Street Hospital, Carmichael House on Aungier Street, Kennedy's Pub on Westland Row, Hewitt's corner, and the Sunlight Building on Essex Quay.

The twentieth century

Concrete buildings built in the last thirty years are usually covered with a thin veneer or cladding of cut stone. Granite and other igneous rocks, from Wicklow or imported from Scandinavia, Brazil and elsewhere, are most common today. This is notable in large buildings, such as the Financial Services Centre where native and imported granites are used side by side. In addition, metamorphic gneiss from Scandinavia and Brazil is coming into vogue. Italian marbles and marble agglomerate (a man-made rock of marble clasts and powder cemented with resin) are now available. Limestone, both native and imported, is also used, as are varieties of travertine.

Cladding tends to be very thin — it is rarely more than two centimetres thick, and is stuck to the building using adhesive, cement, or metal ties or bolts. In many cases around the city, cladding is coming away from the concrete beneath.

Dublin has a rich variety of stone types in its buildings. While the last thirty years have been bleak for the Irish quarries, a recent resurgence in the popularity and use of Irish stone is taking place. Recent reports by the Geological Survey of Ireland (see Bell, 1992 a, b, c, d) indicate some seventy-four granite localities that may be suitable for extraction of reasonable-sized blocks of stone.

Stone such as the limestones from Ballinasloe and Carlow, and Leinster Granite, is available for facing and dimension stone, while old quarries, such as the slate quarries at Killoran, have reopened. Let us hope that this trend continues, and that Irish stone becomes more widely used in our cities and towns.

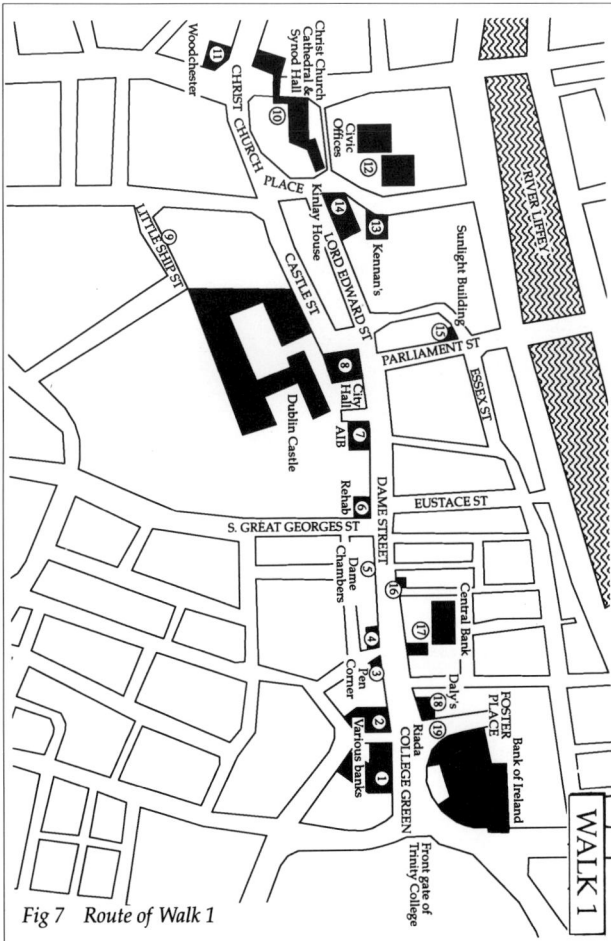

WALK 1

DURATION: 2 HOURS APPROX.

COLLEGE GREEN • DAME STREET
CHRIST CHURCH CATHEDRAL
PARLIAMENT STREET
ESSEX QUAY AND STREET
DAME STREET • COLLEGE GREEN

Fig 7   Route of Walk 1

WALK 1

COLLEGE GREEN (south side)

Start at the front gate of Trinity College and cross to the south side of College Green. This area was once called Hoggen Green (from hauge — *a Viking burial ground).*

1.1 The Bank of Ireland (National Branch) on the left is noteworthy for its decorations, which are carved in granite. They would have been difficult to produce, on account of the coarse texture of the rock, and so they reflect the considerable skills of the stonemasons. It is unusual to find granite so delicately carved.

College Green south. A group of bank buildings, including the fine façade of the Ulster Bank.

The Ulster Bank was built in the 1880s of limestone from Ballinasloe, Co. Galway. The limestone is pale grey in colour, and unfossiliferous (Plate 2a). It is characterised by wavy horizontal lines or stylolites, which mark the site of calcium carbonate solution produced by pressure from overlying sediment. The porch contains panels of Cork Red Marble. The railings are mounted on a dark granodiorite pediment, in which the pale feldspars stand out against the dark basic background.

Further down the street the Ulster Bank extension is faced with Leinster Granite from Ballybrew, near Enniskerry, Co. Wicklow. It is predominantly grey in colour and contrasts well with the pinkish-coloured granite — similar to the variety called Ghiandone Sardo, from Italy — found at the entrance to the bank.

1.2 Cross Trinity Street. On the corner, the National Irish Bank, built in 1879 largely of Irish limestone, has been extended several times. Further down the street is the Allied Irish Bank. It is built of New Red Sandstone, of Triassic age, probably from Cheshire, England (Plate 3d). It has been suggested to me that this stone came from the excavations of the Mersey Tunnel. This rust-coloured stone was deposited in arid desert conditions during the Permian and Triassic geological periods. You will see that the layers are oblique to the horizontal: this is called cross-stratification. Black igneous gabbro called Bon Accord is seen at ground level. This is a fine-grained basic igneous rock, rich in iron and magnesium-bearing minerals, which give it the dark colour. Much of this rock comes from southern Africa, but it is also available in other parts of the world. Note the door surround, which is elegantly carved from the distinctive and common igneous rock larvikite, quarried in Norway. This rock is correctly termed a syenite; it contains large feldspars which are petrol-blue in colour and produce a blue sheen or schiller. It comes in many varieties, ranging in shade from dark to light. Seen here is the former, which is called Emerald Pearl. Next door, the Scottish Legal building is surrounded by a thin band of a paler larvikite, known as Blue Pearl (Plate 1d). Perhaps the finest use of larvikite in Dublin is in St Patrick's Park, near St Patrick's Cathedral, where a large fountain is constructed of Blue Pearl.

1.3 The Pen Corner has a facing at street level of Connemara Marble from Streamstown, Co. Galway. This has not been a success, as the stone has lost its polish in the city atmosphere and has turned black. Such deterioration happens very fast, usually occurring within six weeks. The upper storeys of this building are very attractive. They are highly ornate, with Early Christian symbols carved on the walls, which are surmounted by chimney pots resembling chess pieces.

DAME STREET (south side) to CORK HILL
Dame Street takes its name from the Dam Gate of the City Wall, where a dam controlled the flow of the River Poddle. It was widened by the Wide Streets Commission in 1785/6 as part of a scheme to improve the infrastructure and communications of the city. The Commission widened Westmoreland Street, opened Parliament Street and D'Olier Street, and delineated the South Circular Road, all of which resulted in an ambitious building

programme. This included the erection of buildings such as the Parliament House, City Hall, and Daly's Club, as well as the identical terraced houses of Westmoreland and D'Olier Streets.

In general, modern architecture has made good use of stone and available building materials. However, the street contains one of the worst examples of shoddiness. I refer to the concrete pillars and their capitals on a building close to Dame Chambers. The pillars are capped with half a dustbin lid! The street has seen more prosperous times, but it is now earmarked by the city planners for sensitive revitalisation.

1.4 The building on the corner of Dame Street and St Andrew Street is faced with an English orange calcareous sandstone. Note the balustrade at roof level, which has been painted a pale beige.

1.5 Walk over the mosaic advertising the Stag's Head pub to the entrance of Dame Chambers. This is faced with an attractive deep red rock, Old Gold Granite from Sweden, which takes its colour from the feldspars.

1.6 At the junction of South Great George's Street and Dame Street is situated the imposing building that houses the Rehab Lottery Office. Formerly a Burton's clothing outlet, it is faced with cream-coloured, glazed terracotta tiles called Carraraware, produced for all Burton's outlets by Doulton's of Lambeth. The roof is covered with zinc which has tarnished to a beautiful rich purple. Also used is a pale variety of Emerald Pearl Larvikite, which is often used in banks or for the tops of bars. Next door you will see a dark green rock which is criss-crossed with white veins of quartz. This igneous rock is called serpentinite and comes from Greece, being known in the trade as Verde Tinos.

1.7 The Munster and Leinster Bank (AIB) is built of Ballinasloe Limestone in an Italian baroque style. It resembles the AIB building on the corner of College Street and Westmoreland Street (see Walk 3), both designed by T N Deane in the 1870s.

The pillars are of red granite from Peterhead, near Aberdeen, Scotland. The building was extended about forty years ago, and Ardbraccan Limestone from Co. Meath was used for the walls, and a red Swedish granite for the columns; this stone is slightly darker than that from Peterhead. Look closely at the junction of the two limestones: they do not quite match. Walk past the park,

marked by a commemorative plaque to Dr Barnardo, towards
City Hall.

LORD EDWARD STREET to CHRIST CHURCH PLACE

1.8 City Hall stands on what was once the site of several taverns
or inns, frequented by dandies and notorious for duelling,
gambling and drunkenness. These included the Eagle Tavern,
where the Hell Fire Club was founded in 1735.

City Hall was designed by Thomas Cooly for the city merchants
in 1769, and took ten years to build. It is faced with Portland Stone
(Plate 3b) and capped with a fine patinated-copper green dome.
Examination of the stone facing the River Liffey shows that it has
deteriorated badly. It is blistered in places and supports a good
population of algae and mosses. These plants tend to retain water,
which accelerates the breakdown of the Portland Stone.

Across Cork Hill stands Newcomen's Bank (Rates Department,
Dublin Corporation). Designed in 1781 by Thomas Ivory, it was
one of the few private banks that flourished in eighteenth-century
Dublin. Another was that of the La Touche family, situated
around the corner in Castle Street. If you look carefully you will
see that the swags on either side of the building are of different
colours. This is because the right-hand side and porch were added
later, in 1856–8.

Walk to the gate of Dublin Castle. The fine gate piers are carved
from Portland Stone. The building to your right has recently been
restored. Alternating layers or courses of Ballyknockan Granite
(Plate 1a) and Leacarrow Limestone produce an attractive pattern.
The limestone is bush-hammered, which produces a 'punched'
finish, and the outer edges of each ashlar block are slightly
rebated so as to suggest rustication. The high wall on Castle
Street, with its stucco face, has insets of granite from Three Rock
Mountain, Co. Dublin. This granite is characterised by large, flat,
platy, silvery mica crystals up to one centimetre in width and
three centimetres in length, and large, pale feldspars which are set
into a fine-grained matrix or groundmass dominated by quartz.

Returning to Lord Edward Street you are walking over some of
the best examples of granite paving remaining in Dublin. The kerb
is double-stepped because the drop to street level is quite high,
and the corner paving slabs are laid in a quarter-circle style.
Another example of this is found at the junction of Nassau Street
and South Frederick Street (Walk 2).

A plaque on the wall of Lord Edward Street tells that it was
opened in 1886. Here the considerate planners have put in a

drinking fountain which is carved from Cork Red Marble from Midleton. The outer lip of the basin has been polished through use by many thousands of thirsty Dubliners. It is a pity that it is not in use today.

CHRIST CHURCH PLACE

1.9 Pass the Dublin Castle Inn (the poet James Clarence Mangan was born in a house on this site) and walk down Werburgh Street. Turn left into Little Ship Street, whose name is derived, not from its proximity to the Dubhlinn (Black Pool) where Viking boats were moored, but from a corruption of 'sheep', which once grazed on the nearby pastures. Here a portion of the City Wall remains, together with one of its attendant towers, that of Stanihurst. The City Wall was probably built of Calp Limestone, and surrounded the medieval city. This small stretch was recased and pointed in 1856, and like St Audoen's Arch it bears little resemblance to the original.

Return to Christ Church Place, turn left and walk westwards to the Peace Garden. This park is paved with Liscannor Flagstone, characterised by traces of unknown organisms (Plate 3a). In medieval times, this was the site of the fourteenth-century Tholsel or City Hall. This fine building, depicted by the artist James Malton in 1793, was torn down in 1809.

1.10 Christ Church Cathedral, one of the most interesting buildings in Dublin, was founded in 1032 by Sitric, King of Dublin, who built a wooden church on the site. The present cathedral, which is noted for its crypts, was begun in 1172 by the Normans, whose stonemasons used locally quarried Calp Limestone from Lucan, and an imported oolitic limestone from Dundry Hill near Bristol, which they used around the doors and windows. The remains of the chapter house of the original priory are situated on the southern side of the cathedral site. Here the various rock types may be examined at close quarters using a hand-lens.

The cathedral was restored (actually nearly rebuilt) by the architect George Street in 1878. He used Calp Limestone from Kimmage and Rathgar for the walls, and faced them with limestones from Ardbraccan, Co. Meath, and Sheephouse, near Drogheda, Co. Louth. These latter stones were also used for dressings and carvings. The cathedral was reslated with olive green slates from Eureka in America. However, these were

replaced about twenty years ago with green Westmoreland Slate.
Examples of the American slates remain on the tower and south-
west turret of the Synod Hall. The restoration of Christ Church
Cathedral and the Synod Hall cost £250,000 and was financed by
Sir Henry Roe, a whiskey distiller. Perhaps he wanted to outdo
Benjamin Lee Guinness, the brewer, who had paid for the
restoration of St Patrick's Cathedral in 1864.

Restored portions of the cathedral are easy to pick out. The
blocks are usually ashlared and well pointed. The two most
eastern windows on the south-facing aisle wall have replacement
architraves, which differ in colour from the other windows of the
wall. In 1982 some deteriorated stonework was replaced with
limestone from Leacarrow, near Tullamore, Co. Offaly. ·

1.11 Before leaving Christ Church Place, cross the road to the new
offices on High Street. Is granite used as a facing stone at ground
level? Look closer. It is not granite but pseudo-granite: concrete
containing flecks of mica.

CHRIST CHURCH PLACE to LORD EDWARD STREET (north
side), via FISHAMBLE STREET

1.12 Return to Lord Edward Street. There you can see the Civic
Offices. These are faced with large but thin slabs of Ballyknockan
Granite. The upper slabs are outlined in black, perhaps owing to
water percolation — so beware of walking beneath them!

1.13 Fifty metres down Fishamble Street, on the right-hand side,
stands Kennan's Iron Foundry. This was the site of the Musick
Hall where, on 13 April 1742, Handel's *Messiah* was first
performed. A bronze plaque on the neglected front of the building
marks this important event.

Return up Fishamble Street and turn down Lord Edward Street.

1.14 USIT Kinlay House is an attractive building that until
recently housed the Working Boys' Home and Harding Technical
School, which opened in 1892. It was designed by William
Murray, who was also responsible for Baggot Street Hospital and
Hodges Figgis bookshop on Dawson Street. The combination of
red brick and yellow terracotta facings became known as
'Murray's Mellow Mixture'.

PARLIAMENT STREET to COLLEGE GREEN (north side), via
ESSEX QUAY and STREET and EUSTACE STREET and
DAME STREET

1.15 Walk down into the River Poddle valley, which is situated at
the sharp dip in Dame Street close to the City Hall, and turn left
into Parliament Street. This street was opened in 1762 by the Wide
Streets Commission. Continue on to Capel Street Bridge. On your
left at the corner of Parliament Street and Essex Quay is the
Sunlight Building, with the beautiful enamelled terracotta frieze
telling the story of the manufacture of soap and testifying to the
former use of the building by Lever Brothers.

Return to Dame Street via Essex Street, with the old Dolphin
Hotel on the right — its façade is of carved ornate limestone; and
thence via Eustace Street, with the Irish Film Centre cinemas, the
Friends' Meeting House, and the Shamrock Chambers, decorated
with carved fruit but no shamrock!

Turn right, and a few metres down Dame Street you arrive at
Fan's Chinese restaurant. In building terms this establishment is
notable for the use of an artificial or reconstituted stone called
agglomerate marble. This material consists of ninety-six per cent
Italian marble chips and powder, which have been cemented
together with resin. This material is easily cut, and takes a good
polish. It is cheaper than natural marble, and has an overall
pleasant effect. In this case the variety is called Portoro. Verdi
Alpi, an attractive green variety, is used in the Men's Department
of Brown Thomas in Grafton Street.

1.16 Walk towards Trinity College past Nico's Restaurant and the
unusual leaded glass façade of the building next door. On the
corner of Dame Street and Fownes Street Upper is an insurance
building, of a Gothic-type design. Granite makes up the storey at
street level, and Bath Stone, a Jurassic oolitic limestone, has been
used in the upper storeys. These materials are crossed by
horizontal string courses of New Red Sandstone and grey
limestone. The building carries ornate carvings of wild dogs and
foliage, which enhance the capitals and window architraves. It is
unusual in that the only entrance is the small door opening onto
Fownes Street Upper.

1.17 The controversial Central Bank is faced with Ballyknockan
Granite, while its neighbour, the Commercial Building, is in

Ballybrew Granite. The granite blocks of the latter are flat-faced, with a bevelled, rusticated margin. Close examination of the Ballyknockan Granite reveals its constituent minerals: pale, glassy quartz, milky feldspars, and black and silvery micas. The Ballyknockan variety is medium- to coarse-grained, with interlocking crystals some three to four millimetres in size. Apart from the minerals mentioned above, the granite also contains some chlorite. The Ballybrew Granite is generally coarser grained, containing mica up to ten millimetres in length, smaller quartz and feldspars, and rare tourmaline crystals, which are thin, shiny, elongated black crystals.

Old photographs show that the Commercial Building (built in 1796) once faced Dame Street; when the Central Bank was erected the Commercial Building was turned through ninety degrees. Before demolition of the original, each block was numbered. However, these were not reused, and they now lie in a yard behind Newbridge House near Donabate. The building once housed the Ouzel Galley Society, an arbitration board for commercial interests. The society was formed when a ship that was thought to have been lost at sea in 1705 returned to Dublin laden with booty after an absence of five years. As the insurance had been paid on the boat it was unclear who owned the cargo. Eventually it was agreed by all interested parties to put the assets into the new society. This is commemorated by a medallion on the nearby Dame Street.

COLLEGE GREEN and FOSTER PLACE
1.18 Cross Anglesea Street. Daly's Club once occupied the impressive building that has a fine red clock by Booth in the centre of its façade. Daly's was the most celebrated gambling house in Dublin during the late 1700s. The building was designed by the architect Richard Johnston and erected in 1789. It was connected to the adjacent Parliament Building by an underground corridor, so that the tired legislators could revive their spirits and unload their purses. Some memorabilia, notably the voting box with its black balls, are preserved in the Civic Museum. The building is constructed of Golden Hill Granite: the ground floor is rusticated while the upper storeys are faced in flat ashlar blocks. The attic storey is a more recent addition. The façade bears unusual paired Ionic pilasters. The two wings have long been replaced.

Riada Stockbrokers occupy the site of the east wing of Daly's Club. The building has been attractively refaced in Portland Stone and Westmoreland Green Slate (Plate 4d). Relict bedding may be picked out in the slate, which was originally a volcanic ash before it was subjected to huge pressure and metamorphosed. Examine the smooth unweathered surface of the Portland Stone and compare it with the pitted surface of the adjacent Parliament Building, which demonstrates the effect of two hundred years of weathering.

1.19 Turn into Foster Place. Here the road is made with dark green setts of diorite, a basic, intrusive igneous rock composed of feldspar, hornblende (a green basic mineral) and mica. These came from Charles Stewart Parnell's quarry at Avondale, Co. Wicklow. Parnell, leader of the nineteenth-century Irish Parliamentary Party, employed 200 sett dressers from 1884 until 1891, the year of his death. Without his guidance the business went into decline.

Apart from diorite, many setts are of imported Welsh and Scottish stone. Note the carved items of artillery over what was once the Guard House of the Parliament Building.

College Green is dominated by two buildings. The first, Trinity College, is described in Walk 2. The second is the Bank of Ireland. This was formerly the Parliament Building and was built between 1729 and 1731, to the plans of Edward Lovell Pearce. It has a façade forty-seven metres long with a portico supported by impressive Ionic columns. The building has Calp Limestone rubble walls, which have been faced with Portland Stone. If you look closely at this stone you will find that it is full of shells, predominantly fossils of oyster shells.

The bank building served as a Parliament from its opening in 1731 until the Act of Union in 1800. The Parliament was divided into two chambers: the House of Lords with 120 members, and the House of Commons with 300. In 1804 the building was converted for use by the Bank of Ireland. The Commons chamber was subdivided by Francis Johnston, but the Lords chamber remains unchanged. This room may be visited by arrangement with an attendant of the bank. It contains the mace of the House of Commons and a fireplace of black Kilkenny Marble. The Cash Office is floored in a chessboard pattern with slabs of Cork Red Marble alternating with Portland Stone.

Crossing the street we return to our starting point outside Trinity College.

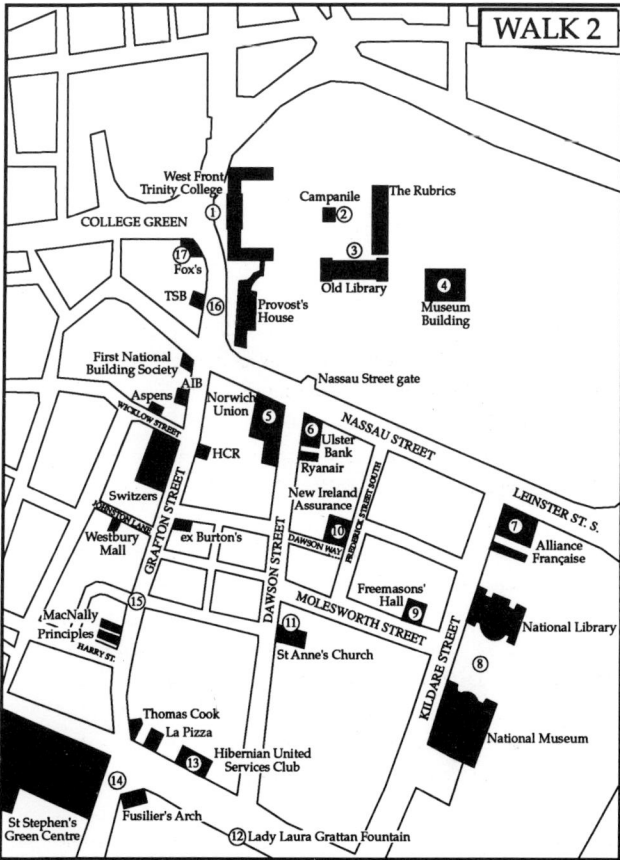

WALK 2

DURATION: 2 HOURS APPROX.

TRINITY COLLEGE
NASSAU STREET • DAWSON STREET
ST STEPHEN'S GREEN NORTH
KILDARE STREET
GRAFTON STREET

Fig 8   Route of Walk 2

COLLEGE GREEN and TRINITY COLLEGE

2.1 Trinity College was founded in 1592 on the site of All Hallows Priory. Many distinguished Irishmen and women have studied in the University, including Wolfe Tone, Robert Emmet, Oliver Goldsmith, Jonathan Swift, Samuel Beckett and Mary Robinson.

The imposing entrance to Trinity College, consisting of a central area flanked by two square pavilions, was built in the 1750s of Leinster Granite from Golden Hill, Co. Wicklow, and Portland Stone was used for the architraves, swags, and Corinthian pilasters and half-columns. The Golden Hill Granite is so called because it has been partially weathered *in situ*, and the slightly altered feldspars give the rock a brownish hue. The masonry cost £27,000. Between 1990 and 1992 the central portion of the building was cleaned. Passing through the gateway one walks over a wooden floor of interlocking hexagonal setts (similar in pattern to the basaltic Giant's Causeway), and into Parliament Square, which is dominated by the identical Corinthian fronts, in Leinster Granite and Portland Stone, of the Chapel on the left and the Examination Hall on the right. Further into the square on the left-hand side is the Dining Hall, restored after a fire in 1984. For reasons unknown, until 1870 the clock in the portico was set fifteen minutes after Dublin time.

Trinity College, Dublin: the Campanile and the Old Library.

2.2 The Campanile between Parliament Square and Library
Square houses the Great Bell of Trinity, which is rung before the
conferring of degrees. It also tolls prior to examinations! This bell
weighs thirty-seven hundredweight and cannot be swung in the
belfry as it is too large. It is rung by chiming instead.

The lower portion of the Campanile is composed of a fine-
grained bluish granite from Blessington, Co. Wicklow, while the
upper cupola is made of Portland Stone.

2.3 Walk across the cobbled square towards the Old Library. The
cobbles are mainly rounded limestone clasts, but they also include
white quartz and the igneous rocks granite, andesite and diorite.
Unlike paving setts, these cobbles are not dressed, but were
probably collected, already rounded, from beaches. They were
relaid during the 1970s.

The Old Library was built between 1712 and 1732 by Col.
Thomas Burgh to house the large college collection of books,
including the *Book of Kells*. The lower storey is built of muddy,
well-bedded Calp Limestone (Plate 2b), cut into regular rusticated
ashlar blocks, which were quarried at Palmerstown, Co. Dublin.
This rock is quite fossiliferous and contains tiny cubic crystals of
iron pyrites or 'fool's gold', a sulphide, which indicates the anoxic
conditions of the Lower Carboniferous seafloor in which it was
deposited. The Calp has weathered to a pleasant, warm, brownish
colour which contrasts well with the grey Ballyknockan Granite of
the upper storeys. Originally these levels were faced with white St
Bee's Sandstone from Whitehaven in Cumbria, but this
disintegrated quickly and all but the carved cornice was replaced.
According to Dr E J McParland of the History of Art Department
in Trinity College, a doorway of Scrabo Stone that formerly was in
the Library is now in Waterford Cathedral.

On the left stand the Rubrics, built of brick in 1700. The terrace
was much altered in 1894 through the addition of Dutch-style
gables, with copings of New Red Sandstone, possibly also from
Scrabo, Co. Down.

2.4 Standing in New Square, one is surrounded by residential
buildings on two sides, built of Golden Hill Granite, by the
Rubrics, and by the Museum Building which was finished in 1857.
This building was designed by the architects Deane and
Woodward, who incorporated Venetian designs popularised in

the 1850s by John Ruskin. It was built at a cost of 25,000 guineas.

The exterior walls of the building are faced with Ballyknockan Granite; the carved, ornate string courses, quoins, columns and capitals are in Portland Stone; the architrave over the heavy wooden door is in Caen Stone (an oolitic limestone from France, favoured by the Normans); and the roof is slated with Welsh 'Bangor Queen' slates. The circular wall veils are unsatisfactorily inlaid with Irish marble, which discoloured in the Dublin atmosphere within ten years.

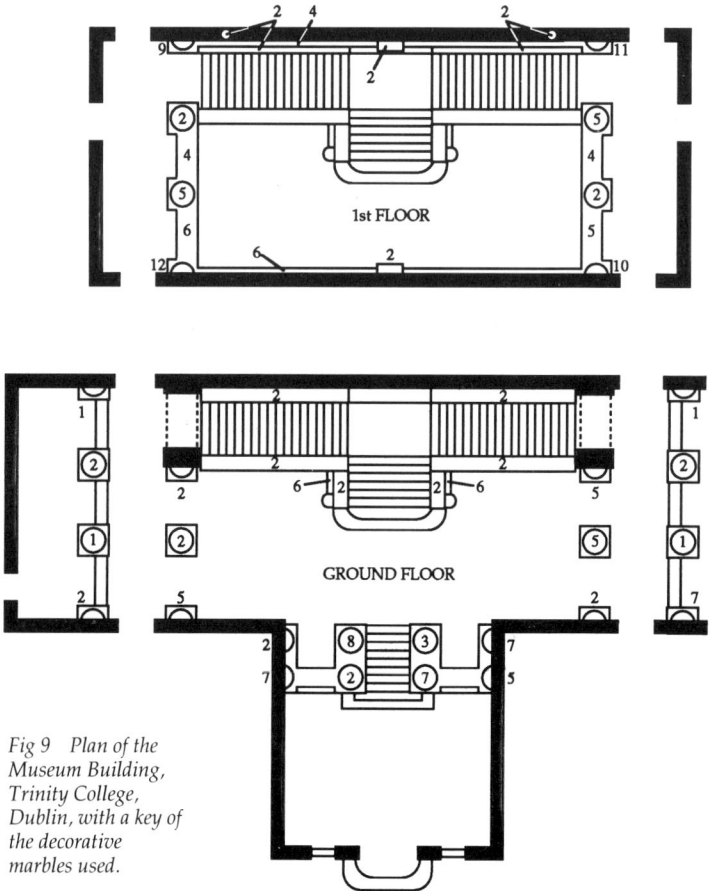

Fig 9   Plan of the Museum Building, Trinity College, Dublin, with a key of the decorative marbles used.

The interior of the building is breathtaking. The large central hall is a geology lesson in itself, and is dominated by a pair of Giant Irish Deer skeletons. Extensive use of colour has been made: the twin-domed roof is composed of blue, red and yellow enamelled bricks; the walls are of creamy yellow Caen Stone; the arches show use of alternating Caen Stone and stained red blocks; the floor is composed of yellow Yorkshire flags interlocking with black slate, bordered with purple Cambrian slate from Pen-yr-Orsedd Quarry, Nantlle, North Wales; and the columns, string courses and banisters are of different Irish marbles. The green marble, which contains the minerals serpentine, chlorite and mica, is the well-known Connemara Marble from Barnaoran (Plate 4a) and Clifden, Co. Galway; the red variety comes from Little Island (Plate 2d), Baneshane and Midleton, three quarries near Cork; the brown columns are from Armagh and Clonony, Co. Offaly (Plate 2c); while the black varieties come from Mitchelstown, Co. Cork, Galway and Kilkenny. The very dark rock that is cross-cut by blood-red veins is a metamorphic serpentinite, which was shipped directly from The Lizard in Cornwall. On the floor-plan of the Museum (Fig 9) the stone types are marked as follows:

1   Lizard Serpentinite. *A distinctive black metamorphic rock composed largely of olivine and pyroxene, with veins of red oxidised serpentine minerals. Locality: The Lizard, Cornwall.*

2   Connemara Marble. *A pale green marble rich in serpentine, chlorite and mica. Resulted from metamorphism of limestone 580 to 475 million years ago. Locality: Barnaoran, Co. Galway.*

3   Connemara Marble. *A dark green variety of this distinctive stone, which shows alternating folded layers of dark and green minerals. Locality: Clifden, Co. Galway.*

4   Baneshane Cork Red Marble. *A red limestone (stained by iron oxides), with blebs of white calcite, and linear concentrations of red clay. Looks like corned beef. Locality: Baneshane, Cork.*

5   Midleton Cork Red Marble. *Similar to Baneshane. This stone was very popular with late-nineteenth-century builders and architects. It was known in the trade as 'Victoria Red'. Locality: Midleton, Co. Cork.*

6   Little Island Cork Red Marble. *A red limestone with patches of white calcite, and small, circular, crinoid ossicles. Much sheared by Hercynian earth-movements. Locality: Little Island, Cork.*

7   Armagh Marble. *A pale, relatively unfossiliferous, brown limestone. Locality: Armagh, Co. Armagh.*

8   Clonony Marble. *A brown limestone with cavities infilled with white calcite; containing many fossils, including crinoids and some cephalopods. Locality: Clonony, Co. Offaly.*

9   Galway Marble. *A black, relatively unfossiliferous limestone. Galway was, and remains, one of the major sources of black marble, which has been worked there since the 1700s. Locality: Menlo Park, Co. Galway.*

10 Mitchelstown Marble. *A black 'reef' limestone containing many cavities infilled with white sparry calcite; these cavities contrast with the black bulk of the rock, which is composed of fine-grained lime mud called micrite. Locality: Mitchelstown, Co. Cork.*

11 Kilkenny Marble. *A black limestone which was much sought for internal work. It is very fossiliferous, containing crinoids, corals and large brachiopods. Locality: South of Kilkenny.*

12 Castle Caldwell Marble. *A pale grey fossiliferous limestone, rich in crinoidal debris. This marble is not well known, and does not appear to have been widely used in the nineteenth century. Locality: Castle Caldwell, near Belleek, Co. Fermanagh.*

The capitals and string courses in both the interior and exterior are heavily carved with representations of flowers and animals. The sharp-eyed may spot cats, snakes, frogs, squirrels and birds, as well as shamrock, daffodils and irises. These carvings were executed *in situ* by the talented O'Shea brothers from Co. Cork, who also worked on the Kildare Street Club for Deane and Woodward.

Leave Trinity through the Nassau Street Gate via the Arts Building.

NASSAU STREET to MOLESWORTH STREET, via KILDARE STREET

2.5 Cross Nassau Street and walk towards Grafton Street for a few metres. On your left the entrance to the Norwich Irish Building Society on Nassau Street is faced with Sardinian Grey Granite. The modern building on the corner (Norwich Union Corner Development), and its neighbours on Nassau Street (Nassau House) and Dawson Street (Norwich Union House), are supported at ground level by square pillars, which are clad in thin panels of imported igneous rock. Nassau House and the Norwich Union Corner Development are faced with Bon Accord stone — a dark gabbroic rock, often called Black Granite in the stone trade. This basic igneous rock contains considerable amounts of dark feldspars, pyroxene, hornblende and olivine. Varieties of this stone are available in India, Sweden and South Africa.

Walk up Dawson Street to Norwich Union House, which has pillars clad in Balmoral Red Granite (Plate 1b), imported from Finland. Balmoral Red Granite got its name because much of it was imported through Aberdeen after the Scottish sources were exhausted. Next door, the entrance to the Norwich Union has been faced with flame-textured Pink Porrino Granite (Plate 1c) from Spain, containing diagnostic clumps of pink feldspars, and with Leinster Granite from Ballyknockan.

2.6 Cross the road to the Ryanair office. There the doorway is paved with black slabs of dark limestone, which contain large white brachiopod shells. This easily recognisable limestone is from Paulstown, Co. Kilkenny. Walk back towards Nassau Street. The building on the corner, occupied by the Ulster Bank, is built of Irish fossiliferous limestone. It stands on the site of Morrison's Hotel, which was frequented by Charles Stewart Parnell.

2.7 Walk down Nassau Street to the Alliance Française, on the corner of Nassau Street and Kildare Street. This building, built largely of brick, once housed the Kildare Street Club. The pillars and some string courses are of a bluish limestone quarried at Ballyhule, near Tullamore, while other string courses are of Portland Stone. Note the splendid carved animals, especially the billiard-playing monkeys and the hare being chased by greyhounds, executed in Portland Stone by C W Harrison of Great Brunswick (Pearse) Street.

A few doors up there is a grey, austere building, the only redeeming feature of which is that it is faced with a very fossiliferous limestone from Carlow. The rock is festooned with fossils: crinoids (like the cross-section of bolts), brachiopods (shells) and corals. Carlow Stone has also been used in the recently erected façade of Newmount House on Lower Mount Street.

2.8 The National Museum and the National Library are reached further along Kildare Street. These buildings, erected in 1890, were faced with Leinster Granite, while buff-coloured micaceous Mount Charles Sandstone from Donegal was used on the upper storeys and for dressings around doors and windows. Over 3000 tons (3048 tonnes) of the sandstone were transported to Dublin and dressed on site.

The director of the Geological Survey of the time said of this rock that 'any mineralogist will predict for it remarkable freedom from decay'. Sadly, he was quite wrong. The sandstone has been badly affected by Dublin's coal-polluted atmosphere, and that on the National Library was replaced in the 1960s by a grey limestone from Ardbraccan, Co. Meath. Some replacement of stone was necessitated as early as 1932. Close examination of the stone remaining on the National Museum shows obvious decay and exfoliation of the outer layers of the rock, caused by the breakdown of the ferrous cement used to bind the sand grains together. It is interesting to note that this rock has not been broken down by the atmosphere in its native Co. Donegal, and survives there in a number of buildings dating from as early as 1820. Replacement of badly eroded stone by Leinster Granite is currently continuing on the National Museum.

2.9 In Molesworth Street stand some of Dublin's few remaining 'Dutch Billy' houses, with their distinctive gables. The

Freemasons' Hall, dating from 1866, is built of Ancaster Stone, a Jurassic creamy-buff coloured limestone from Lincolnshire, England.

SOUTH FREDERICK STREET to DAWSON STREET, via DAWSON WAY

2.10 Turn right down South Frederick Street and walk to the New Ireland Assurance Building. It was refaced in 1988 with granites which show two different finishes. The red granite, Carmen Red from Finland, has been flame-textured, which tends to reduce the colour intensity of the large, red, blotchy feldspars characteristic of the rock. The Mondariz Granite from Sweden, inserted in the window architrave, has a honed or sawn finish, and is characterised by the development of very large feldspars, which show a regular tabular, crystal shape.

Inside the foyer of the building, polished Carmen Red Granite and Sardinian Grey Granite have been used to create an open, bright reception area. The three granites used in this building were shipped from their Scandinavian and Sardinian sources, were cut and finished in northern Spain, and were finally shipped to Ireland for fitting!

Dawson Way is faced in Carmen Red Granite and Sardinian Grey Granite, and is floored with Leinster Granite.

Walk to Dawson Street via Dawson Way and turn left towards St Stephen's Green.

2.11 Half way up on the left is St Anne's Church, which is faced with Ballyknockan Granite. The stark greyness of this stone has been broken by horizontal string courses of Irish limestone, pale Portland Stone, and New Red Sandstone from the Triassic of Scotland or Liverpool. This elegant façade was added by Thomas N Deane in 1868. His proposed second tower was not executed.

ST STEPHEN'S GREEN north to GRAFTON STREET

2.12 The Lady Laura Grattan Fountain at the junction of Dawson Street and St Stephen's Green was built in 1879 of granite from Cornwall. This distinctive rock contains large tabular, rectangular, aligned feldspar crystals. The horse trough on the left-hand side (as you look down Dawson Street) is also of Cornish granite, possibly from Halvossa, whereas the trough on the right is made of Leinster Granite. The fountain was restored in 1992. Lady

Laura Grattan was the wife of Sir John Grattan, a soldier who served at Waterloo under Wellington, and the eldest son of the parliamentarian Henry Grattan.

2.13 The Hibernian United Services Club, the façade of which has been cleaned recently, is faced with brick from Dolphin's Barn, Dublin. The Club occupies the town house built for the Bishop of Killala, which was added to by A E Murray in the 1890s. Further along the street, just beyond the former Galleria, the building occupied by La Pizza is faced at street-level with Baltic Brown Granite from Finland. This stone is similar to the Carmen Red Granite from the same country, in that the feldspar crystals are large and circular in shape, but it differs from it in that the crystals are dark brown in colour.

The entrance to St Stephen's Green is dominated by the Fusiliers' Arch. This monument bears the names of Irishmen who fell in the Boer War of 1899–1900. It is built of Ballyknockan Granite, and the pale limestone insets are from Sheephouse, near Drogheda.

2.14 The recently built St Stephen's Green Centre commands a prominent position on the corner of South King Street and St Stephen's Green. It is faced in Barnacullia Granite from Co. Dublin. This granite is similar to that from Ballybrew, except that it contains large micas fifteen to twenty millimetres in length, which appear like crayon marks in cut surfaces. Inside, the shopping malls are floored with tiles of Arabescato Fantastico, a white Italian marble, together with a dark grey marble. Unfortunately these rock-types are quite soft, unlike granite, and have become damaged in places by stiletto heels. The entrance and steps to Hickey's fabric shop are of Norwegian Pink Marble (also called Fauske Marble), which is quarried close to the Arctic Circle, and is not pink but pale orange in colour.

GRAFTON STREET to COLLEGE GREEN

2.15 Walk down Grafton Street. In this busy street there are some unusual facing stones, and excellent use of Irish stone. Thomas Cook, on the left-hand side, is faced with Maracana Gneiss from Brazil (a similar stone is used on Cook's office in Lower Grafton Street, opposite Trinity College), (Plate 4c). This reddish-black rock has been heavily altered: it is folded and its minerals have become differentially layered. On the front of Principles clothes shop, just past the Harry Street junction, a beautiful salmon pink New Red Sandstone, with well developed stratification or bedding, has, sadly, been recently painted a rust colour (to match stone on the upper storeys). Around the door a pink granite containing huge feldspar accumulations (similar to those seen in Dawson Lane) is used. A few doors down, Red Brazilian Granite may be seen at MacNally, the opticians. This granite is characterised by the deep red colour of the feldspar crystals.

Make a detour down Johnston Lane and go to the Westbury Mall. The floor is paved in Burren Stone, a mica-bearing flagstone from Co. Clare, similar to the better-known Liscannor Flagstone. Burren Stone contains vertical worm burrows (or trace fossils) which lie perpendicular to bedding and have the deceptive appearance of raindrop marks, while in the Liscannor variety the traces are largely horizontal to bedding.

Retrace your steps to Grafton Street and turn left. The former Burtons' building on the junction with Duke Street has the same cream-glazed Carraraware terracotta tiles as seen earlier in Dame Street. Further down the street, Hayes Conyngham and Robinson's premises are faced at ground level with a greenish-pink granite from Galway. Across the street, Switzers' façade sports white marble, probably from Sicily, and the Allied Irish Bank has recently been refaced using the dark grey fossiliferous Ballinasloe Limestone from Top Quarry, Co. Galway. At the junction of Grafton Street and Suffolk Street, the First National Building Society has been faced using grey granite as well as the distinctive blue larvikite from Scandinavia. This igneous rock, in which the blue feldspars dominate, is more strictly called a syenite. The interior of the First National Building Society has been completely clad with Roman Travertine (Plate 3c), a creamy, highly porous sedimentary rock which is precipitated by the evaporation of water at hot springs in Italy. Note the irregular holes which have been infilled with plaster. A red variety of this rock is found on the premises of Aspens Bar and Restaurant in Wicklow Street.

2.16 Walking towards Trinity College, you will see the Provost's House on your right-hand side. It was begun in 1759 and is the last remaining large mansion in residential use in the city centre. It is built of Golden Hill Granite and Ardbraccan Limestone, which cost £2830. Some authors claim that red sandstone from

Liverpool was used, but examination of this building cannot substantiate this.

As can be seen from the walls of the house, the granite has suffered quite badly from atmospheric pollution in the last two hundred years. In places it has been covered by a sand-cement mortar which has the appearance of sandstone. The limestone blocks have been carved with icicle-like patterns. This form of decoration is called frost-work.

Opposite the Provost's House stands the Trustee Savings Bank, housed in a splendid building. The facing stone at street level is dark polished gabbro, above which is salmon-pink to purple-coloured New Red Sandstone on the first floor and a dark yellow-orange, iron-bearing, calcareous sandstone on the second floor. Bedding or stratification can be clearly picked out in this Scottish sandstone. Fine columns of red granite, perhaps the classic Peterhead Granite from Scotland, protect the entrance to the main trading hall. Inside this hall are four marble Corinthian columns. However, tapping and examination of these shows them to be made of coloured plaster — an effective substitute for real marble! This is called scagliola or 'scag'. Pillars of this type are easily recognised as they are often four to six metres high, and appear to be made of one piece of stone.

2.17 Fox's tobacconist is housed in a fine Gothic-style Victorian building, recently cleaned, with some interesting carved heads. The rock is an English, fine-grained, micaceous, calcareous sandstone, probably of Triassic age. Clearly visible on the ashlar blocks are the tooling marks made by the masons' chisels.

The walk has now returned to its starting point outside Trinity College.

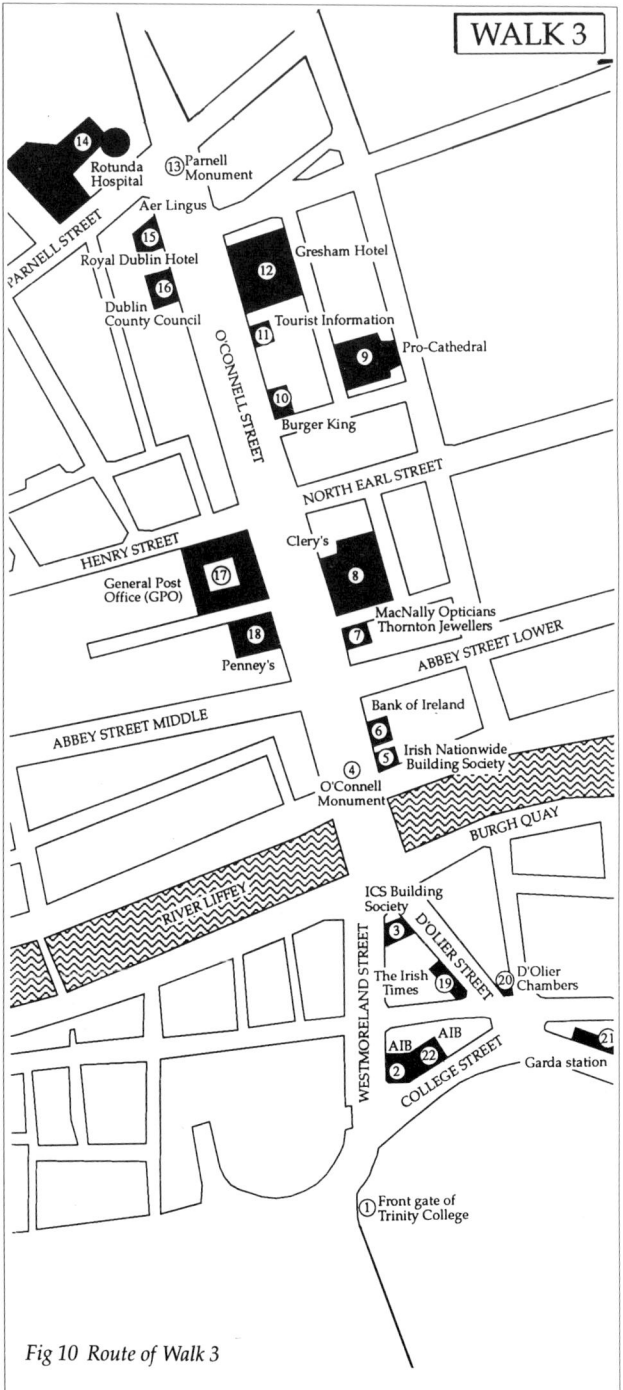

Fig 10 Route of Walk 3

DURATION: 2 HOURS APPROX.

WESTMORELAND STREET
O'CONNELL STREET
MARLBOROUGH STREET
D'OLIER STREET • PEARSE STREET

COLLEGE GREEN to O'CONNELL BRIDGE

3.1 Start at Trinity College. Cross College Street and stop by the statue of Thomas Moore. For a short time after it was erected the statue was the subject of ridicule from passers-by. It was referred to as 'Black-a-Moore', and mothers would chide their children with the warning that Moore would get them!

Unfortunately Moore and his statue were short and stumpy. To enhance the work and its presence the statue was decapitated and over seven centimetres of neck added. Set into the pavement beside the adjacent public convenience is a plaque commemorating the Dublin Millennium of 1988; it quotes a line from Joyce's *Ulysses* referring to the Dubliner's call of nature.

3.2 Cross College Street to the building on the corner with Westmoreland Street. It is built of New Red Sandstone, either from Scrabo, Co. Down, or from Liverpool. The building contains an attractive mix of colour, with the grey limestone pillars and white Portland plaques highlighting the red sandstone of the walls. Notice the difference in colour between this stone and that of the liver-red Scottish sandstone from Dumfries of the buildings on either side. Westmoreland Street once contained two rows of standard buildings designed by Aaron Baker for the Wide Streets Commission in 1800. However, these stretches are broken on both sides of the street by several newer buildings, including the Educational Building Society with its façade of glass and granite.

3.3 The ICS Building Society is faced with a number of granites, including native Ballyknockan Granite and Pink Porrino Granite,

41

which have a rough-textured natural surface. Cross the traffic
island with the embedded bronze footprints to O'Connell Bridge.

It is unfortunate that the view down towards the Custom House
and the mouth of the Liffey has been spoiled by Liberty Hall, the
new Financial Services Centre, and the Loop Line Bridge which
now carries the DART (Dublin Area Rapid Transit) between Tara
Street and Connolly (Amiens Street) stations. This metal bridge
was completed in 1891. The piers are constructed of granite from
Dublin and Aughrim, Co. Wicklow, and of Carboniferous
limestone from Tullamore, Cavan and Meath. They are decorated
with brick from north Devon and Chester in England. Butt Bridge,
close by, is made of Ballyknockan Granite.

O'CONNELL STREET (east side)

Most of the buildings in this, the main thoroughfare of the city,
were destroyed by shelling during the 1916 Rebellion. Shells were
fired from a gunboat on the River Liffey. This ship, the *Helga*, was
formerly the research vessel of the Department of Agriculture and
Technical Instruction. Rebuilding took place soon after 1916. It is
unfortunate that the planners did not take the opportunity to
implement an overall plan for the street. Instead they allowed
piecemeal reconstruction, which gives it an untidy aspect.

*O'Connell Street, with the monument to Daniel O'Connell on the left
and the General Post Office in the right background.*

3.4 The O'Connell Monument which dominates the lower end of O'Connell Street was erected between 1864 and 1882. The grey limestone plinth is surmounted by a 3.6 metre bronze of Daniel O'Connell, 'The Liberator'. Beneath him is a frieze of some fifty figures, each signifying a different worthy theme: Erin is facing the river and holds the Act of Emancipation in her left hand. The whole monument is protected by winged figures on the corners, representing Patriotism (holding a sword and shield), Fidelity (with an Irish Wolfhound), Eloquence (clutching a book), and Justice (with a serpent).

Just up the street stands the statue of Sir John Gray, who was responsible for the Vartry Waterworks which improved the city's water supply. The statue is executed in a white Italian marble from Serravezza.

3.5 Cross Eden Quay to the Irish Nationwide Building Society. Here panels of Blue Pearl Larvikite (syenite), with its distinctive petrol-blue colour, are found. Just down the street the darker variety of this rock, Emerald Pearl, may been seen at Hamilton Long, the camera shop.

3.6 Further along the street the Bank of Ireland (the former Hibernian Bank) is built of Leinster Granite and Portland Stone. The façade is notable for its fine, but unusual, fluted columns cut in Leinster Granite. At Broadway Amusements, panels of particularly striking Blue Pearl Larvikite are to be found.

3.7 Cross Abbey Street Lower. At Thorntons the jewellers and MacNally the opticians the black, basic igneous rock gabbro has been used to good effect.

3.8 The building occupied by Clery's is architecturally the third best building on the street, after the General Post Office and the Gresham Hotel, and is faced with Portland Stone. It was built soon after 1916 and consists of a steel frame onto which the stonework was attached. Note the fine carving, in particular the fluted columns, and the abundant clam and oyster shells.

NORTH EARL STREET to MARLBOROUGH STREET, and back to O'CONNELL STREET via CATHEDRAL STREET
Turn right into North Earl Street with its new statue of James Joyce, and walk to Marlborough Street junction. Turn left and continue to the Pro-Cathedral.

3.9 The Pro-Cathedral was begun in 1815 to the design of John Sweetman. Sweetman was in exile in France at the time because of his involvement with the 1798 Rebellion, and he submitted a scale-model of the building. The front is executed in a Grecian-Doric style, with columns and portico in Portland Stone. This stone was soaked in oil in an attempt to make it less susceptible to weathering. It is difficult to assess the value of this action, as non-treated stone of the same age would have to be examined. Return to O'Connell Street via Cathedral Street and turn right.

3.10 The building occupied by Burger King is not remarkable. However, it is worth looking at, because at ground-floor level four different stone types have been used. These are all currently available and are typical of stone used in the last two decades. The pale granite is from Barnacullia, Co. Dublin; the black stone is gabbro; the red granite is the Swedish Balmoral Red; while above the plate glass windows the cream-coloured Portland Stone is used.

3.11 Stop outside the Tourist Information Office. Here another method has been attempted to stop or retard deterioration of Portland Stone. In this case a clear silicone coating has been spread over the surface of the fluted columns. Its effect is awful. The texture of the stone is lost and the silicone has turned a pale green colour.

3.12 The Gresham Hotel is regarded as one of the finest buildings on the street. Like Clery's, it is faced with Portland Stone which is highly carved.

PARNELL STREET
3.13 The Parnell Monument was erected in 1911 to commemorate Charles Stewart Parnell. It is built of two Galway granites. Around the monument the bollards and the slabs between the

The Rotunda Hospital and the Parnell Monument.

cobbles are of Barna Granite. This stone is quite dark in colour, and contains large pink feldspars (called phenocrysts) which are set into a fine-grained, greenish-black matrix of quartz, feldspar and mica. The obelisk is in Shantalla Granite, finer grained, with pink and green feldspar crystals interspersed with small quartz and micas. The bronze of Parnell is by the Irish-born sculptor Augustus St Gaudens.

■● ●■● ●
  ●● ●●

3.14 The Rotunda or Dublin Lying-In Hospital was built in 1757, and still functions as a maternity hospital. The walls are faced with Leinster Granite and Kilgobbin Granite. Later additions to the complex included the Rotunda or round room and the buildings now occupied by the Gate Theatre. The former building was executed in Portland Stone and Leinster Granite, to which a sculptured frieze of ox heads and other panels were added. These are interesting as they are made of Coade Stone, a fashionable artificial stone used widely in the late 1700s. This material was also used for the panels and urns adorning the Rutland Fountain opposite the National Gallery on Merrion Square.

O'CONNELL STREET (west side)
3.15 The Aer Lingus office is faced at ground-floor level with Portland Stone, into which string courses of Leinster Granite have been set. Above is a reddish-black metamorphic rock called

gneiss, which probably comes from Sweden. This rock may originally have been a granite which was drastically altered by heat and pressure generated by burial and earth movements. It is also used next door at the Royal Dublin Hotel, where a deep red granite may be examined.

3.16 Outside the office of Dublin County Council, flat slabs of pale Roman Travertine have been laid. You will notice that dirt has penetrated the pores of this stone, and has picked out the bedding. The use of travertine as a paving stone is not successful, and it is better employed indoors. This stone is also used at the Dublin Bus office, further down the street, where an Irish grey limestone has been finished in an unusual and attractive way. The flat surface of the stone has been decorated with a grid-pattern of grooves.

3.17 The General Post Office (GPO) was built during the years 1824 to 1826 to the design of Francis Johnston, at a cost of £50,000. Its Ionic portico, with fluted columns, is faced with Portland Stone, while the side wings are in Leinster Granite.

It was gutted during the 1916 Rebellion. In 1928, the interior was refitted, and much use was made of Connemara Marble and black fossiliferous Kilkenny and Galway marbles. At the same time the roof was reslated using Irish slate from Killaloe, Co. Clare, which has weathered to a green patina. Of note inside the main trading hall is the statue 'The Fall of Cú Chulainn' by Oliver Sheppard.

3.18 Penney's department store occupies a building that displays on its upper frontage the contrasting colours of blue-grey Ballinasloe Limestone from Co. Galway and paler igneous Ballybrew Granite from Co. Wicklow. The ground-floor level is clad with Sardinian Grey Granite, Pink Porrino Granite and Balmoral Red Granite.

O'CONNELL BRIDGE to COLLEGE GREEN, via D'OLIER STREET and COLLEGE STREET

Cross O'Connell Bridge and proceed down the right-hand side of D'Olier Street. On the west side stands a terrace of standard buildings erected by the Wide Streets Commission. They are probably similar if not identical to those that once stood around the corner in Westmoreland Street. Opposite, note the 1930s façade of the Gas Company, with its splendid stained and frosted glass depicting gas workers. Next door the Blessed Sacrament Chapel is faced with a honed Irish grey limestone from Ballinasloe, which is offset nicely with pilasters of Balmoral Red Granite.

3.19 Near the junction with Fleet Street, the terrace occupied by *The Irish Times* was restored in 1991, when the original glass and granite façade was reinstated.

3.20 Cross D'Olier Street to Pearse Street. You will pass by D'Olier Chambers, a fine Victorian terracotta and brick building.

3.21 Make a quick detour up Pearse Street, which will bring you to the Garda Station. This attractive grey building is faced with Leinster Granite rock-faced blocks from Ballybrew, which have been laid in a random manner. Note the carved heads of the policemen over either side of the doorways.

Retrace your steps and continue down College Street towards the east portico of the Bank of Ireland. The final building on this walk is the Allied Irish Bank on your left-hand side.

3.22 This was the Provincial Bank when it was constructed in 1865. The portico is in Bath Stone. Inside the banking concourse, the ceiling is decorated with fine plasterwork. Tall pillars of red Peterhead Granite emphasise the size of this huge room.

Return to your starting point outside Trinity College. Have a cup of tea.

Government Buildings, Upper Merrion Street.

St Stephen's Green east, with the Office of Public Works on the left. The hallway of this building contains forty panels of Irish decorative marbles.

WALK 4

DURATION: 2 HOURS APPROX.

LINCOLN PLACE
MERRION STREET LOWER
MERRION SQUARE WEST
MERRION STREET UPPER • MERRION ROW
ST STEPHEN'S GREEN EAST, SOUTH
AND WEST

This walk passes along three sides of St Stephen's Green, as well as through an area of Dublin that developed after the north-side residential streets and squares such as Henrietta Street and Mountjoy Square were laid out. The south side became fashionable soon after the Earl of Kildare built his residence, Leinster House (now the Oireachtas or Houses of Parliament), in 1745. Merrion Square and adjacent streets followed after 1762.

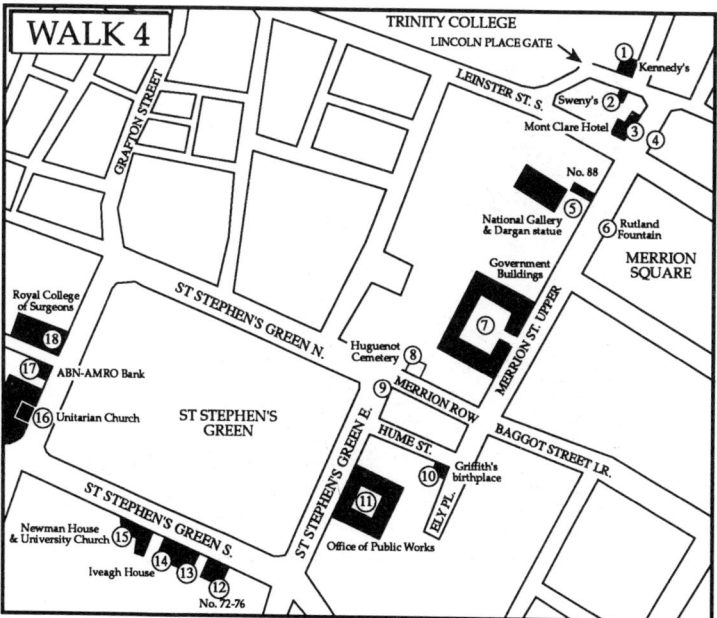

WALK 4

TRINITY COLLEGE
LINCOLN PLACE GATE
① Kennedy's
LEINSTER ST. S.
Sweny's ②
Mont Clare Hotel ③ ④
No. 88
National Gallery & Dargan statue ⑤
Rutland Fountain ⑥
GRAFTON STREET
MERRION SQUARE
Government Buildings
ST STEPHEN'S GREEN N.
Royal College of Surgeons ⑱
⑰ ABN-AMRO Bank
⑯ Unitarian Church
ST STEPHEN'S GREEN
Huguenot Cemetery ⑧
⑦
⑨ MERRION ROW
MERRION ST. UPPER
BAGGOT STREET LR.
HUME ST.
ST STEPHEN'S GREEN E.
Griffith's birthplace ⑩
⑪
ELY PL.
Office of Public Works
ST STEPHEN'S GREEN S.
Newman House & University Church ⑮
Iveagh House ⑭ ⑬ ⑫
No. 72-76

Fig 11 Route of Walk 4

LINCOLN PLACE to MERRION SQUARE

The walk commences at the Lincoln Place Gate of Trinity College, beside the Dental Hospital. Opposite, across the street, stands a modern office block on the site of the old Turkish Baths. Walk along Lincoln Place to the junction with Westland Row.

4.1 The façade of Kennedy's Pub is a mixture of pebble-dash interspersed with elaborate terracotta tiles. In particular note the decorative tiling at roof level along Westland Row. Perhaps when you have finished this walk, you might care to return for a leisurely drink in comfortable and convivial surroundings.

Oscar Wilde was born in a house on Westland Row, and beyond is Pearse Station, from where the first train journey made in Ireland set off on 17 December 1834.

4.2 Cross the road to Sweny's Pharmacy, which is mentioned in James Joyce's *Ulysses* and which is noted for its Lemon Soap; why not go in and buy a bar, and carry it in your pocket as Bloom did? Merrion Hall, which was burnt down in 1992, has been converted into a hotel, the Davenport, and only the façade remains of the original building.

4.3 Moving around the corner, walk to the new entrance of the Mont Clare Hotel. Here the pavement and kerb have been newly relaid using granite from Co. Wicklow. This represents a welcome change in policy, as not so long ago, old granite was being removed from the city's pavements and replaced with concrete slabs. Apart from being more attractive, the granite offers better grip to pedestrians, particularly when it is raining.

Just inside the door of the modern extension to the Mont Clare Hotel, four different rock types can be examined and contrasted. The steps are of Leinster Granite, matching that of the pavement; polished tiles of an orange-black streaky rock cover the floor of the foyer, which is surrounded by a thin strip of black gabbroic rock, while the junction of the floor and the wall is marked by the use of Balmoral Red Granite from Scandinavia. The orange-black rock is a metamorphic rock called a gneiss, from Sweden, which shows differentiation into poorly defined layers of dense (dark) and lighter (pale) minerals. Originally this rock was a granite, which was altered by great heat and pressure that occurred during orogenic (mountain-building) periods over 500 million years ago.

∎◗ ▗◗ ▝◗ ◗▖

4.4 On the opposite side of the street, beside Sir William Wilde's house, Yorkshire Flagstones are used as paving slabs. There are few examples of these honey-coloured micaceous sandstone slabs remaining in Dublin; some have survived in steps outside the headquarters of the Fine Gael party at 51 Upper Mount Street.

MERRION SQUARE west

4.5 Cross over Clare Street and walk towards the National Gallery. Stop at the doorway of house number 88 and examine the columns of Ballinasloe pale grey fossiliferous limestone. They are covered in places with deposits of soot. However, the resistance of this stone to deterioration in Dublin's atmosphere contrasts sharply with the heavily eroded state of the Portland Stone used in the carved capitals and bases of the columns.

The National Gallery houses an important collection of European and Irish art, and benefits from royalties bequeathed to it by George Bernard Shaw, whose statue stands in its grounds. The dominant statue situated in the centre of the front lawn is that of William Dargan (1799–1867), a wealthy philanthropist and railway enthusiast who lived close by in Fitzwilliam Square. He developed the Dublin to Kingstown Railway in 1834 and organised the Great Exhibition of 1853 which took place on Leinster Lawn. While Dargan is of interest, it is the plinth that deserves our more immediate attention. Walk over the grass and have a close look at the rock. It contains medium-sized crystals of green and pink feldspars, as well as clear quartz. It is an example of the Galway granite that has already been encountered at the Parnell Monument on O'Connell Street. However, this rock is unusual. Unlike most building stones that are quarried, this rock was a huge glacial erratic, found at Ballagh, Co. Roscommon. It had been carried there by ice from west Galway, during the last Ice Age, and perched on top of Carboniferous limestone until it was removed for use under Dargan. You will also see a small block of red Peterhead Granite, from Aberdeen, on which Dargan's left elbow is comfortably resting.

∎◗ ▗◗ ▝◗ ◗▖

4.6 Cross over the street to the Rutland Fountain. This was designed by H Aaron Baker, who was responsible for

Westmoreland and D'Olier Streets, and it was erected in 1791. Decorating the fountain are a numbers of urns, as well as oblong and round plaques with figures and profiles of heads, all of which are cast out of Coade Stone. One of the urns once adorned the house on the corner of Clare Street and Merrion Square. The walls of the fountain are of Leinster Granite and the fluted string courses are of Portland Stone.

Return to the west side of the street and continue towards Government Buildings, passing the Natural History Museum on your right as you go.

MERRION STREET UPPER

4.7 The impressive white building that forms most of the west side of Merrion Street Upper houses Government offices. This large building with its impressive front courtyard was designed by Sir Aston Webb and Sir Thomas Manley Deane (grandson of Sir Thomas Deane of Kildare Street and Trinity College Museum Building fame), and was opened in 1911 for use by the Royal College of Science for Ireland. By 1926 this institution was amalgamated with University College Dublin, which until 1990 retained the central portion of the building for its School of Engineering. Since that time the whole building has undergone considerable refurbishment and all the façades have been cleaned.

This was the last major Dublin building to have been largely faced with Portland Stone. A close look at the creamy-white stone in some of the pillars reveals horizontal layers or beds of numerous shelly fossils, mainly oysters, that thrived in shallow, warm seas in southern Britain some 150 million years ago. The fossils tend to stand a few millimetres proud of the rock surface. This is because they are slightly more resistant to weathering, and no doubt to the cleaning processes, which tend to remove some of the surrounding rock. Some blocks of stone contain cavities; these are moulds of fossils such as gastropods. In recent times this cavernous rock, called Portland Roach, has been favoured for use in London, particularly as it is less expensive than the less fossiliferous variety from the Whit bed at Portland. The basement level is faced with Ballyknockan Granite, in which schlieren or accumulations of mica can easily be seen.

Opposite Government Buildings stands a fine row of Georgian houses. The finest of these is Mornington House, where the Duke of Wellington was born in 1769. It was until recently occupied by the Land Commission.

MERRION ROW to ST STEPHEN'S GREEN

4.8 Walk around the corner to Merrion Row and proceed to the Huguenot Cemetery. This was one of the burial places of Protestant exiles who fled France from the 1600s onwards. A plaque of grey slate, mounted on the gate pier, tells that the cemetery was cleaned up and restored in 1988 by FÁS (The Industrial Training Authority) with the assistance of the French Ministry for Foreign Affairs. The use of slate for this plaque was a good choice as it weathers very slowly and the quality of the carved lettering remains good for many years.

Many gravestones that date from the mid 1700s and the 1800s are made of slate. This is fortunate for those interested in tracing genealogies, as many of the inscriptions can still be read after two hundred years. This contrasts sharply with lettering cut into granite, which is illegible after about a hundred years. Slate headstones are rarely seen in modern graveyards, where Italian marble, imported and native granites, and Irish limestone are the stones most commonly used. For those of you who seek a degree of immortality, a slate headstone, or as a second choice, one carved from an Irish limestone, should ensure that your name will remain on view for several centuries to come!

ST STEPHEN'S GREEN

St Stephen's Green is an area of trees, lawns, shrubbery and duck ponds which has been enjoyed by Dubliners for many, many years. In medieval times it was a common, used for grazing stock, and its boundaries spread beyond those with which we are familiar today. During the seventeenth century, the area became run down and was a haunt for petty thieves and prostitutes. It became very fashionable from the early 1700s, when members of the nobility and some clerics built impressive town houses there. These include Iveagh House and Newman House. In the early 1800s, Dublin Corporation erected the present railings, and then rented the park to the occupants of the houses that surrounded it. Lord Ardilaun (son of Benjamin Lee Guinness who paid for the restoration of St Patrick's Cathedral) purchased the lease on the land in 1880 and presented it to the people of Dublin. It has been open to the public ever since. The ponds in the park were laid out at this time, and are fed with water piped from the Grand Canal at Portobello.

Today St Stephen's Green contains memorials to a number of Irishmen and women, including Theobald Wolfe Tone, O'Donovan Rossa, Countess Markievicz, James Joyce, William

Butler Yeats, and the victims of the Great Famine of the 1840s.

The street landscape of St Stephen's Green has suffered in the last three decades through the demolition of several buildings. Many of the smaller houses on the southern and western sides are now gone. Only Iveagh House, Newman House, the University Church, on the south, and the Unitarian Church and the Royal College of Surgeons on the west, remain unscathed, together with a handful of smaller houses. The rest have been replaced by modern office blocks and by the large shopping centre near the Grafton Street junction. A number of houses on the eastern side are recent additions, built in a Georgian style.

4.9 Cross Merrion Row. There, at the north-east corner of St Stephen's Green, stands a fine piece of sculpture. It consists of two arches set at right angles to each other, and is built of Portland Stone, Irish fossiliferous limestone (probably from Carlow), and bronze. Erected during the Millennium celebrations of 1988, it was inspired by Georgian Dublin architecture, and cleverly mimics arches and keystones reminiscent of the Custom House. Note the dark and light stripes running from the arches to the pavement kerbstones. What stone do they remind you of? Liscannor Flagstone with its trace fossils? If you look closely you will see that each piece is identical — they are made of concrete. In any case Liscannor stone is dark grey and not white in colour. Why a poor imitation was used when the real thing is readily available is a mystery to me.

Pass the Virginia creeper-covered house on the left, which is one of the most attractive houses remaining on the Green. Continue on and turn into Hume Street.

HUME STREET
4.10 Beyond Hume Street Hospital, at the junction of Ely Place, stands the house in which Sir Richard Griffith was born in 1784. Griffith, who died at the advanced age of ninety-four years, was a geologist who produced the first large-scale geological map of the complete island in 1839 and who was also responsible for the Griffith Valuation of property surveyed between 1830 and 1868. His birthplace is marked by a fine plaque of Ballinasloe Limestone, with lettering by Michael Biggs, the foremost Irish stone engraver of the last thirty years.

ST STEPHEN'S GREEN east

Return to St Stephen's Green and walk towards Leeson Street. If you are not careful you might bypass, without a thought, a building that has an important past and is a rich source of information about Irish marbles of the mid nineteenth century.

4.11 Number 51 St Stephen's Green, now the headquarters of the Office of Public Works, contained from 1846 the Museum of Economic Geology (later the Museum of Irish Industry), an institution established by the chemist Sir Robert Kane (1809–90), as well as the Geological Survey of Ireland. In 1867, the Museum of Irish Industry was incorporated into the Royal College of Science for Ireland, which was also housed in the building, while the Geological Survey moved around the corner to 14 Hume Street in 1870. The building was finally vacated by the Royal College of Science for Ireland when it moved to splendid new premises on Upper Merrion Street in 1911 (see 4.7 above).

The foyer of number 51 contains forty panels of Irish marbles. These were put in place in March 1850 by Kane, whose book *The Industrial Resources of Ireland* (1844) brought to wider public notice the range and variety of Irish marbles. Much of Kane's information was taken from George Wilkinson's book *Practical Geology and Ancient Architecture of Ireland* (1845), which described the physical properties of over six hundred Irish stone types. The Great Exhibition of 1853 was another factor that resulted in the development of a valuable marble industry in Ireland in the second half of the nineteenth century.

The marble panels are of twenty-six stone types, which display a range in colour from black, to red, brown and green. With the exception of the Connemara Marble, all the 'marbles' are Lower Carboniferous limestone, which take a good polish. These limestones, which underlie nearly fifty per cent of the country, were deposited in a warm, shallow sea that migrated northwards across Ireland between 360 and 333 million years ago, and which contained a rich fauna of cephalopods, bryozoans, corals, brachiopods and fish. The remains of some of these animals may be seen in a number of the 'marbles', particularly the Kilkenny black variety.

The stone types contained in the forty panels are numbered as follows:

1	Cork Grey Marble: *'reef' limestone, pale grey in colour, veined with stromatactis.*
2	Cork Red Marble: *red limestone containing blebs of paler calcite and crinoidal debris. Probably from Little Island, Cork.*
3, 23, 28, 29, 37	Kilkenny Black Marble: *black, largely unfossiliferous limestone. It may contain few small solitary corals and other fossil debris.*
4	Scawt Hill (Scaughthill) Marble, Co. Antrim: *pale yellow/fawn-coloured stylolitic stone, with pale, irregular blotches, ten centimetres in diameter, randomly distributed. Metamorphosed Cretaceous chalk.*
5	Clonmacnoise Pink Marble, Co. Offaly: *pale grey limestone containing stromatactis filled with pinkish calcite.*
6, 7	Kilkenny Black Brachiopod Marble: *contains circular sections of brachiopod shells, six centimetres in diameter, which are set in a matrix of black lime mud.*
8	Pallaskenry Grey Marble, Co. Limerick: *dark grey crinoidal limestone with paler indistinct stromatactis.*
9, 11, 13	Armagh Brown Marble: *pale, relatively unfossiliferous, brown limestone.*
10	Clonony Grey Marble, Co. Offaly: *grey-brown goniatite-bearing stylolitic stone.*
12	Armagh Grey Marble: *mottled dark grey stone crossed by veins of white calcite.*
14	Connemara Marble: *(light variety) sepia in colour, probably from Ballinahinch, Co. Galway.*
15	Mitchelstown Marble, Co. Cork: *black 'reef' limestone which contains many cavities infilled with white sparry calcite, which contrast with the black micritic bulk of the rock.*
16	Clonony Brown Marble, Co. Offaly: *brown limestone with cavities infilled with white calcite; containing many fossils, including crinoids and some cephalopods.*
17	Connemara Marble *from Ballinahinch, Co. Galway: sepia in colour.*
18, 32	Pallaskenry Purple Marble, Co. Limerick: *reddish purple colour, with finely communited, randomly distributed crinoidal fragments.*
19, 25	Churchtown Red Marble, Co. Cork: *red limestone, stained by iron oxides, with irregular patches of white calcite.*
20	Connemara Marble *from Clifden, Co. Galway: streaked, foliated variety with distinct sepia and black layering.*
21, 24	Galway Black Marble: *unfossiliferous dark limestone.*
22	Galway Black Coral Marble: *dark limestone containing large colonies of* Lithostrotion *coral.*
26	Clonmacnoise Grey Marble, Co. Offaly: *pale grey fossiliferous limestone containing a large volume of debris from fossil crinoids or sea lily skeletons.*
27, 30	Cork Red Marble: *red brecciated limestone containing blebs of paler calcite. Probably from Midleton, Co. Cork.*
31	Carrigaline Marble, Co. Cork: *pale grey, moderately fossiliferous limestone, containing crinoids and corals.*
33, 36	Churchtown Grey Marble, Co. Cork: *grey, unfossiliferous, 'reef' limestone, with stromatactis cavities filled with white calcite.*
34	Doneraile Marble, Co. Cork: *black, unfossiliferous stone.*
35	Limerick Marble: *black stone with an occasional gastropod shell and cross-cut by white veins of calcite.*
38, 40	Monkstown Marble, Co. Cork: *grey, reefal, brecciated limestone containing few fossils and pale stromatactis.*
39	Cork Red Marble: *similar to that from Little Island (No. 2) but containing no crinoidal debris. Probably from Baneshane, Co. Cork.*

After leaving the colourful hall of the Office of Public Works, turn left and walk to the junction with Leeson Street; cross it and Earlsfort Terrace to the south side of the Green. This was once known as Leeson's Walk.

ST STEPHEN'S GREEN south

4.12 You will notice that the first three buildings are large office blocks. The third, number 72–76 St Stephen's Green, is worthy of attention, if only for some of the stones used around its entrance. The steps are fabricated from a pale yellow micaceous sandstone, probably from Yorkshire. The surface of the stone is not always totally flat: you can see areas that are a millimetre or two lower than the upper surface. This is called parting lineation and is caused by the fracture along which the rock is split jumping from one bedding plane to another. It generally occurs in rocks that contain platy minerals such as mica, which become aligned parallel to bedding as the rock is flattened by pressure from above.

Between the steps and the windows and door of the building, slabs of a buff-coloured brecciated marble can be seen. This stone is called Carnic Grey Marble and was quarried in northern Italy. The stone is a heavily stylolitised and fragmented marble and is cross-cut by veins of white calcite. The marble was subjected to great stress as the Alps were thrust upwards when the African continental plate jostled with that of Europe over 50 million years ago. This pressure initially caused the rock to fragment into small irregular pieces. This process is called brecciation. Further pressure resulted in solution of the stone, which occurred along irregular planes called stylolites. Insoluble clays in the limestone are now seen as dark irregular lines occurring at these undulating lines. The use of this pale marble around the entrance was a mistake. With use, and exposure to the elements, it has lost its polish, and the minute cracks that permeate the stone have become infilled with dirt. Compare the state of the stone inside and outside the front door. Clearly it should only be used indoors.

The square pillars are faced with Portland Stone, while the wall of the adjoining house and those inside the reception area are faced with a pale Italian marble called Arni Fantastic from Tuscany. The stone in question is marked by thin streaks of green running through the rock, which makes it resemble some blue cheeses. Inside the building the reception counter is faced with a dark green, serpentine-rich marble from Greece called Larissa Green.

4.13 A few doors down the street is a red-brick building occupied by the Department of Foreign Affairs. Here New Red Sandstone, from Merseyside, has been used for window sills, string courses, keystones above the window openings, and a low wall in front of the building. You will be able to make out the cross-stratification in the rock, and you may find some small ripples if you look closely.

4.14 Next door is Iveagh House, numbers 80 and 81 St Stephen's Green, which is also used by the Department of Foreign Affairs. Number 80 was built in 1736 as the town house of the Bishop of Cork. In 1856 Benjamin Lee Guinness purchased and rebuilt it, incorporating the adjacent house. The façade of this four-storey house is faced with blocks of Bath Stone; these are rusticated on the ground levels and are plain ashlar on the upper two floors. Surrounding the roof level is a simple balustrade which increases the elevation of the house.

The steps of the house were originally of Portland or Bath Stone, but were replaced recently by slabs of quartzite from Donegal (Plate 4b). This metamorphic rock was once a sandstone but it underwent recrystallisation during the Caledonian orogeny or mountain-building episode over 400 million years ago. It is a pale white/light green stone speckled with tiny black spots. It breaks easily along bedding planes into thin layers, is strong, and is ideally suited as a paving stone. The Sugar Loaf Mountains in Co. Wicklow are also composed of quartzite, and are not, contrary to popular belief, ancient volcanoes. The Wicklow rock is heavily jointed, which renders it useless for building and paving purposes.

The interior of the house is breathtaking, and contains many fine Italian marble fireplaces. Unfortunately, it is not open to the general public.

4.15 Newman House, which occupies numbers 84 and 85, originally housed the Catholic University (which later became University College Dublin). This institution of learning was founded by Cardinal Newman, and was to cater for those who were denied a university education because of a ban on their attendance at Trinity College Dublin, which was placed by the Catholic hierarchy. The columns outside the doorway are turned in Irish grey limestone. Above the door lies a splendid docile lion, which looks as if it has eaten too many zebras.

According to Maurice Craig (1982), the University Church, next door, was designed by J H Pollen, on the instructions of Newman. This church dates from 1856 and is in a Byzantine style, with attractive alternating bands of red and black brick used in the porch. The ornate capitals and the tympanum over the entrance are of Portland Stone.

The remaining stretch of this side contains rather uninteresting buildings. The most notable is the former Methodist Centenary Church which was built 150 years ago. The church and its site cost £10,000. The church took its name from the then recent celebration of the centenary of Methodism in 1839. It was gutted by fire in 1961 and now, of the original building, only the façade remains, behind which is found an office building.

ST STEPHEN'S GREEN west
Although this side of the Green has been extensively rebuilt in the last ten years, it is dominated, architecturally speaking, by the Royal College of Surgeons.

4.16 The first building of note to be encountered as one walks towards Grafton Street is the Unitarian Church, erected in 1863. This building is dwarfed by the modern bulk of the Banque Nationale de Paris and Telecom Éireann building which is wrapped closely around it. It reminds one of St Patrick's Cathedral in New York, which is similarly crowded by tall buildings, although they are probably ten times higher than their Dublin equivalents! The Unitarian Church presents a fine granite-faced frontage onto the Green. Architraves around the doorways and windows are carved from a soft, yellow, shelly limestone, called Purbeck Stone, which is rather decayed and which comes from the south of England. The buttresses are faced with an Irish grey limestone. If you examine the left-hand side wall of the building you will notice that it is composed of hard, flinty Calp Limestone.

4.17 Walk towards Grafton Street and stop at ABN-AMRO bank. This modern building is faced with a thin cladding of Ballybrew Granite. Of particular interest here is the variety of stone finishes that can be seen. Horizontally laid slabs of cut, unpolished or honed stone surround the flower beds in front of the bank.

Between the pavement and the upper level, slabs of stone that have been flame textured may be examined. Run your hand over the surface: it is quite uneven. This finish is produced by passing a very hot flame over the cut surface of the stone, which causes the crystals of quartz and feldspar to melt slightly. The darkish accumulations of mica may be seen more easily in these slabs than in their cut neighbours. At the base of the wall, some small granite setts have been inserted into the footpath. These display another finish described as 'rock-faced'.

4.18 The Royal College of Surgeons dates from 1806. It was designed by Edward Parke, who was architect to the Royal Dublin Society, and was extended twenty years later by William Murray. The walls probably contain a rubble core of Calp Limestone. They are faced with ashlared and rusticated blocks of grey Leinster Granite, very likely from Golden Hill, Co. Wicklow. The statues above the pediment represent, as one might expect, Health and Medicine. A small plaque on the right-hand side of the building notes that this was a site of much fighting during the Rising of 1916. The small group of Irish troops was commanded by Countess Markievicz.

You have now reached the corner at the top of Grafton Street, where this walk ends. If you have not already undertaken Walk 1, turn to page 37 and continue to Trinity College.

GLOSSARY OF GEOLOGICAL AND ARCHITECTURAL TERMS

ACID STONE: an igneous rock (such as granite) that contains more than sixty-five per cent silica. Usually pale in colour.

ARCHITRAVE: the surround of a window or doorway.

ASHLAR: squared stone blocks with flat faces.

BAROQUE: a flamboyant architectural style commonly associated with the late 1700s.

BASIC STONE: an igneous rock (such as gabbro) that contains less than fifty-five per cent silica. Usually dark in colour.

BEDDING: layering in sedimentary rocks (cf. stratification).

BRACHIOPOD: a bivalved marine animal with two unequal-sized shells. Quite abundant in rocks of Palaeozoic age, but rare today.

BRECCIA: a coarse sedimentary rock consisting of angular pieces (greater than two millimetres in diameter) set into a fine matrix. May be produced by movement of rock along a fault.

BRYOZOAN: a small marine animal (approximately 0.1 millimetres in diameter) which forms skeletal colonies, often in the form of a meshwork. The individual animals feed by trapping food in a tentacle crown. Common in Palaeozoic rocks, especially those of Carboniferous age.

BUSH-HAMMERED: a rock finish where the surface is heavily pitted. It is produced by a hammer that resembles a tenderising hammer used by butchers on tough steak.

CAPITAL: the top of a column, which is usually carved.

CARBONIFEROUS: geological period 360 to 290 million years ago, during which, in the earlier part, Ireland was covered by warm, shallow seas in which lime mud was precipitated and deposited. This became lithified into limestone.

CEPHALOPOD: marine molluscs such as squid, octopus, cuttlefish and nautilus, in which the foot is modified into tentacles around the mouth.

CHERT: a material, generally black in colour, composed of tiny crystals of silica. It is often found occurring in bands within the Calp Limestone of Co. Dublin.

CLADDING: thin slices of rock fixed to the surface of buildings.

CLASTIC ROCK: a sedimentary rock composed of fragments of broken rock that range in size from cobbles to fine mud.

COBBLES: rounded stones used for paving and decorative purposes.

COPING: horizontal course of stone, capping walls.

CRETACEOUS: geological period 144 to 65 million years ago. Characterised by deposition of chalk.

CRINOID: an echinoderm, with a stem, a cup-shaped body, and five or more arms. Commonly called a sea lily. The fragmented portions of the stem (resembling bolts) are most frequently preserved in Carboniferous limestone.

CUPOLA: a dome forming the roof of a building.

DEVONIAN: geological period 408 to 360 million years ago, during which sandstones were deposited on land or in a shallow sea in the Munster region. Old Red Sandstone of Devonian age from Co. Offaly is used locally as a facing stone.

DIMENSION STONE: stone that is available in large blocks or slabs for building or facing.

DRESSED STONE: stone that has been cut or chiselled into regular blocks by stonemasons. Available in several finishes, such as bush-hammered, flamed, honed or rock-faced.

FELDSPAR: a silica-rich mineral (such as orthoclase and plagioclase) containing potassium and aluminium.

FLAME TEXTURED: a finish, applied by heating with a flame, to a flat, cut surface of an igneous rock (generally granite) which results in the surface becoming slightly uneven.

FRIEZE: a decorated panel inserted into the façade of a building.

FROST-WORK: a form of ornamentation, used in the eighteenth century, which resembles icicles.

GABBRO: a coarse- to fine-grained, dark, basic (poor in silica) igneous rock, rich in iron and magnesium-bearing minerals which impart the dark colour.

GASTROPOD: a snail (mollusc).

GNEISS: a highly metamorphosed rock consisting of bands of light and dark minerals.

GONIATITE: a cephalopod mollusc, similar to the modern day nautilus, common in Upper Carboniferous times.

GRANITE: a coarse-grained, pale-coloured, acid (rich in silica) igneous rock composed of the minerals quartz, feldspar and mica.

GRANODIORITE: similar to granite but usually darker.

HONED: a finish given to a rock surface that has been cut and lightly polished.

IGNEOUS ROCK: crystalline rock derived from molten magma, for example, andesite, basalt, diorite, gabbro, granite, granodiorite.

LIMESTONE: a sedimentary rock composed of calcium carbonate ($CaCO_3$), consisting mainly of shell debris and lime mud. Oolitic limestone is composed of tiny spherical particles of calcium carbonate which was precipitated around a nucleus in shallow water.

MAGMA: molten rock derived from deep within the earth's crust or upper mantle.

MARBLE: a metamorphosed limestone, although the term is also used by the stone trade to refer to a limestone that can take a polish.

METAMORPHIC ROCK: a rock that has undergone physical change or metamorphosis through the action of heat and/or pressure on pre-existing rock, for example, marble, slate.

MICA: a silica-bearing mineral (such as biotite and muscovite), containing aluminium or iron or both, which forms flat platy crystals.

OOLITIC: a term referring to limestones that are composed of tiny spherical accumulations (ooids) of calcium carbonate.

OROGENY: a period of mountain building.

PEDIMENT: a triangular structure above a portico, door or window.

PILASTER: a square half-column attached to the wall of a building.

PORTICO: a series of columns along the front of a building.

QUARTZ: a silica-bearing glassy mineral (SiO_2) which is common in many rocks.

QUOIN: the corner of a building, often finished with ashlared or decorated stone.

RELICT BEDDING: in a metamorphic rock, where traces of the original bedding of the sediment can be seen (often in slate).

ROCK-FACED: outer visible portion of a block of stone which has been finished so as to leave the natural broken surface.

RUSTICATION: ashlar blocks surrounded by a carved, sunken or bevelled recess which produces a shadow. Banded rustication is a variety where only the horizontal joints are recessed.

SCHIST: a metamorphic rock that has been formed through regional metamorphism. They are generally foliated, which is produced by the parallel alignment of mica and other minerals within the rock.

SEDIMENTARY ROCK: rock derived from deposited sediments — clastic or precipitated, usually layered or bedded, and often fossiliferous, for example, limestone, sandstone, travertine.

SETTS: regular blocks of stone used for paving.

SLATE: a metamorphic rock produced by the action of pressure on fine-grained sediments.

STRATIFICATION: layering or bedding in sedimentary rocks. Oblique beds or cross-stratification may develop in sand dunes or sediments deposited by rivers.

STRING COURSE: horizontal moulding on the face of a building.

STROMATACTIS: irregular hollows in limestone, filled with white calcite.

STUCCO: ornate plaster work, usually found in eighteenth-century houses.

STYLOLITIC: a limestone containing irregular lines, along which solution of the rock took place when it was subjected to pressure.

SWAG: a decorative motif in the form of draped material, used on some eighteenth-century buildings.

SYENITE: a coarse igneous rock rich in feldspars and poor in quartz, for example, larvikite, a variety from Norway.

TERRACOTTA: a mixture of reddish clay and sand, which may be moulded into tiles, then fired in a kiln and used as a facing material.

TRACE FOSSILS: preserved marks which indicate activity by some organism, rather than the remains of the organism itself, for example, burrows, tracks.

TRAVERTINE: a rock formed by the deposition of calcium carbonate in hot springs found in volcanic regions.

TRIASSIC: geological period 250 to 210 million years ago, when dry terrestrial conditions prevailed in the British Isles.

TYMPANUM: a triangular area above a doorway.

INDEX OF STONE TYPES

BIBLIOGRAPHY

ANON. 1986. 'Stone in Ireland.' *Stone Industries*, June 1986, 20–29.

BELL, A. 1992a. 'The Galway Granites: dimension stone potential.' *Geological Survey of Ireland Report Series* RS92/1.

1992b. 'The Mayo-Sligo Granites: dimension stone potential.' *Geological Survey of Ireland Report Series* RS92/2.

1992c. 'The Donegal Granites: dimension stone potential.' *Geological Survey of Ireland Report Series* RS92/3.

1992d. 'The Leinster Granites: dimension stone potential.' *Geological Survey of Ireland Report Series* RS92/4.

BROAD, I and ROSNEY, B. 1982. *Medieval Dublin: two historic walks.* O'Brien Press, Dublin.

BRÜCK, P M and O'CONNOR, P J. 1977. 'The Leinster Batholith: geology and geochemistry of the northern units.' *Geological Survey of Ireland Bulletin* 2, 107–141.

CRAIG, M. 1952 (reprinted 1980 and 1992 with minor additions). *Dublin 1650–1950.* Allen Figgis, Dublin. [Contains an invaluable bibliography.]

1982. *The Architecture of Ireland from the earliest times to 1880.* Batsford, London and Easons, Dublin.

DAVIES. A C. 1977. 'Roofing Belfast and Dublin 1896–98: American penetration of the Irish market for Welsh slate.' *Irish Economic and Social History* 4, 26–35.

FREESTONE, I C. 1991. 'Forgotten but not lost: the secret of Coade Stone.' *Proceedings of the Geologists' Association* 102, 135–138.

KELLY, A. 1990. *Mrs Coade's Stone.* Self Publishing Association Ltd, Worcs. [Contains a comprehensive list of Coade in Ireland.]

KINAHAN, G K. 1886–89. *Economic Geology of Ireland* — a comprehensive account of all economic aspects of Irish geology of the 1880s. This volume first appeared as separate papers in the *Scientific Proceedings of the Royal Dublin Society*, and was republished complete as Volume 18 of the *Journal of the Royal Geological Society of Ireland*.

HOLDSWORTH, J. 1859. *Geology, Minerals, Mines and Soils of Ireland.* Houlston and Wright, London.

HOLLAND, C H (editor). 1981. *A Geology of Ireland.* Scottish Academic Press, Edinburgh. [The most up-to-date account of Irish geology.]

1987. *Irish Scenery.* Irish Heritage Series, Easons, Dublin.

McNULTY, M. 1991. *Stone, Beautiful Stone: a guide to the skill of practical masonry.* Privately published.

MAX, M D. 1985. 'Connemara marble and the industry based upon it.' *Geological Survey of Ireland Report Series* RS 85/2, 1–32.

MITCHELL, G F. 1986. *The Shell Guide to Reading the Irish Landscape.* Michael Joseph, London and Country House, Dublin.

ROBINSON, E. 1990 and 1991. *The Building Stones of London.* Books 1 & 2. Scottish Academic Press, Edinburgh.

RUCH, J. 1970. 'Coade Stone in Ireland.' *Quarterly Bulletin of the Irish Georgian Society* 13, 1–12.

RYAN, N M. 1992. *Sparkling Granite.* Stone Publishing, Dublin. [Describes the history of granite quarrying in the Three Rock Mountain area of Co. Dublin.]

WHITTEN, D G with BROOKS, J R V. 1983. *A Dictionary of Geology.* Penguin, London.

WHITTOW, J B. 1974. *Geology and Scenery in Ireland.* Penguin, London.

WILKINSON, G. 1845. *Practical Geology and Ancient Architecture of Ireland.* John Murray, London. [Lists stone quarries open in the 1840s.]